Holt Mathematics

Chapter 7 Resource Book

HOLT, RINEHART AND WINSTON
A Harcourt Education Company
Orlando • Austin • New York • San Diego • London

Copyright © by Holt, Rinehart and Winston

All rights reserved. No part of this publication may be reproduced or transmitted in any form or by any means, electronic or mechanical, including photocopy, recording, or any information storage and retrieval system, without permission in writing from the publisher.

Teachers using HOLT MATHEMATICS may photocopy complete pages in sufficient quantities for classroom use only and not for resale.

Printed in the United States of America

If you have received these materials as examination copies free of charge, Holt, Rinehart and Winston retains title to the materials and they may not be resold. Resale of examination copies is strictly prohibited and is illegal.

Possession of this publication in print format does not entitle users to convert this publication, or any portion of it, into electronic format.

ISBN 0-03-078303-8

7 8 9   170   10 09

CONTENTS

Blackline Masters

Parent Letter	1
Lesson 7-1 Practice A, B, C	3
Lesson 7-1 Reteach	6
Lesson 7-1 Challenge	8
Lesson 7-1 Problem Solving	9
Lesson 7-1 Reading Strategies	10
Lesson 7-1 Puzzles, Twisters & Teasers	11
Lesson 7-2 Practice A, B, C	12
Lesson 7-2 Reteach	15
Lesson 7-2 Challenge	17
Lesson 7-2 Problem Solving	18
Lesson 7-2 Reading Strategies	19
Lesson 7-2 Puzzles, Twisters & Teasers	20
Lesson 7-3 Practice A, B, C	21
Lesson 7-3 Reteach	24
Lesson 7-3 Challenge	26
Lesson 7-3 Problem Solving	27
Lesson 7-3 Reading Strategies	28
Lesson 7-3 Puzzles, Twisters & Teasers	29
Lesson 7-4 Practice A, B, C	30
Lesson 7-4 Reteach	33
Lesson 7-4 Challenge	34
Lesson 7-4 Problem Solving	35
Lesson 7-4 Reading Strategies	36
Lesson 7-4 Puzzles, Twisters, & Teasers	37
Lesson 7-5 Practice A, B, C	38
Lesson 7-5 Reteach	41
Lesson 7-5 Challenge	42
Lesson 7-5 Problem Solving	43
Lesson 7-5 Reading Strategies	44
Lesson 7-5 Puzzles, Twisters & Teasers	45
Lesson 7-6 Practice A, B, C	46
Lesson 7-6 Reteach	49
Lesson 7-6 Challenge	50
Lesson 7-6 Problem Solving	51
Lesson 7-6 Reading Strategies	52
Lesson 7-6 Puzzles, Twisters & Teasers	53
Lesson 7-7 Practice A, B, C	54
Lesson 7-7 Reteach	57
Lesson 7-7 Challenge	58
Lesson 7-7 Problem Solving	59
Lesson 7-7 Reading Strategies	60
Lesson 7-7 Puzzles, Twisters & Teasers	61
Lesson 7-8 Practice A, B, C	62
Lesson 7-8 Reteach	65
Lesson 7-8 Challenge	66
Lesson 7-8 Problem Solving	67
Lesson 7-8 Reading Strategies	68
Lesson 7-8 Puzzles, Twisters & Teasers	69
Lesson 7-9 Practice A, B, C	70
Lesson 7-9 Reteach	73
Lesson 7-9 Challenge	74
Lesson 7-9 Problem Solving	75
Lesson 7-9 Reading Strategies	76
Lesson 7-9 Puzzles, Twisters & Teasers	77
Lesson 7-10 Practice A, B, C	78
Lesson 7-10 Reteach	81
Lesson 7-10 Challenge	82
Lesson 7-10 Problem Solving	83
Lesson 7-10 Reading Strategies	84
Lesson 7-10 Puzzles, Twisters & Teasers	85
Answers to Blackline Masters	86

Date _____

Dear Family,

In this chapter your child will learn different ways to organize and display data, about trends and relations in graphs, and about interpreting misleading graphs. Organizing data will help your child use and understand information.

Your child will learn how to quantify data, and how to organize and display it in different ways. The **mean** is the sum of the data values in a data set divided by the number of data items. The **median** is the middle value of an odd number of items arranged in order. For an even number of items, the median is the average of the two middle values. The **mode** is the value or values that occur most often. When all the data values occur the same number of times, there is no mode. The **range,** or the difference between the least and greatest values, is used to show the spread of the data in a data set.

Other ways to display and compare data include frequency tables, stem-and-leaf plots, line plots, bar graphs, histograms, circle graphs, box-and-whisker plots, line graphs, and scatter plots.

A **frequency table** shows how many times an item, a number, or a range of numbers occurs in a set of data.

A **stem-and-leaf plot** shows each piece of data in a set. It allows you to easily identify the median and the mode, as well as the greatest and least value in a set of data. Another way to show this kind of information is a **line plot**.

Stem-and-Leaf Plot

Test Scores

Stems	Leaves
7	8 9 9
8	0 1 1 3 3 3 6
9	0

Key: 7 | 9 means 79

Line Plot

Test Scores

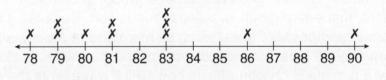

A **bar graph** can be used to display and compare data. The scale of a bar graph should include all the data values. It should also be easily divided into equal intervals.

Holt Mathematics

A **histogram** is a bar graph that shows the frequency of data within equal intervals. It contains no space between the bars.

A **circle graph,** sometimes called a pie chart, shows how a set of data is divided into parts. The entire circle contains 100% of the data. Each **sector,** or slice, of the circle represents one part of the entire data set.

A **box-and-whisker plot** is used to show distribution of the data in a data set. The data is divided into four equal parts, or **quartiles,** in a box-and-whisker plot.

A **line graph** uses line segments to connect data points for different times. The result is a visual record of change over time.

A **scatter plot** helps determine relationships between different data sets. For instance, two sets of data, such as the length and the weight of dinosaurs, may be related. To find out, make a scatter plot of the data values in each set. A scatter plot has two number lines, called *axes*—one for each set of data values. Each point on the scatter plot represents a pair of data values. These points often appear to be scattered.

Your child will learn how to choose an appropriate way to display data, and how to identify **misleading data,** which helps them learn critical thinking skills. For example, advertisements and news articles often use data to support a point, sometimes representing data in a way that influences how the reader interprets the data. A data display that distorts information in order to persuade can be misleading.

Your child will also learn about **populations** and **samples**. When information is being gathered about a group, such as all registered voters, that entire group is called a population. Because it can be difficult or impossible to research all members of a population, we often choose a part of a population, called a sample, to study. The way a sample is chosen affects how well it represents the entire population. A **random sample**, in which each member of the population has an equal chance of being chosen, better represents than does a **biased sample**.

For additional resources, visit go.hrw.com and enter the keyword MS7 Parent.

Name _____ Date _____ Class _____

LESSON 7-1 Practice A
Frequency Tables, Stem-and-Leaf Plots, and Line Plots

The table shows normal monthly temperatures for Tampa, Florida, for each month of the year.

Normal Monthly Temperatures in Tampa

Month	January	February	March	April	May	June
Temp. (°F)	61	63	67	72	78	82

Month	July	August	September	October	November	December
Temp. (°F)	83	83	82	76	69	63

1. Complete the cumulative frequency table for the data.

Normal Monthly Temperatures in Tampa

Temperature (°F)	Frequency	Cumulative Frequency
60–69		
70–79		
80–89		

2. How many months had a normal temperature of less than 80°? _____

3. Complete a stem-and-leaf plot of the data.

Normal Monthly Temperatures in Tampa

Stems	Leaves
6	
7	
8	

Key:

4. How many months had a normal temperature in the 60s? _____

5. Complete a line plot of the data.

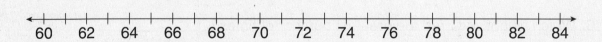

Name _____ Date _____ Class _____

LESSON 7-1 Practice B
Frequency Tables, Stem-and-Leaf Plots, and Line Plots

The table shows the heights of students in Ms. Blaire's class.
Use the table for Exercises 1 and 2.

Height (in.)	
Males	60, 45, 48, 57, 62, 59, 57, 60, 56, 58, 61, 52, 55
Females	49, 52, 56, 48, 51, 60, 47, 53, 55, 58, 54

1. Make a cumulative frequency table of the data.

Heights of Students

Height (in.)	Frequency	Cumulative Frequency

2. How many of the students were less than 60 in tall? _____

3. Make a stem-and-leaf plot of the data.

Height of Students

Stem	Leaves

Key:

4. How many of the students were less than 50 in tall? _____

5. Make a line plot of the data.

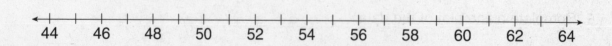

6. Which height occurred the greatest number of times? _____

Name _____ Date _____ Class _____

LESSON 7-1 Practice C
Frequency Tables, Stem-and-Leaf Plots, and Line Plots

The list below shows the ages of all the presidents inaugurated in the twentieth century.

42, 51, 56, 55, 51, 54, 51, 60, 62, 43, 55, 56, 61, 52, 69, 64, 46

1. Make a cumulative frequency table of the data.

2. Make a stem-and-leaf plot of the data.

Key:

3. Make a line plot of the data.

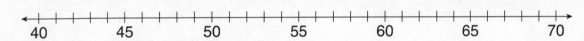

4. How many of these presidents were in their 50s when they took office? _____

5. Which of the following is the most likely source of the data in the stem-and-leaf plot? Explain your answer.

   a. class sizes in a middle school

   b. speeds in miles per hour of cars on a highway

   c. number of hours worked in a week by part-time employees of Toy Town

Stems	Leaves
5	1 1 2 2 3 4 5 6 8 8
6	0 0 0 0 4 1 8
7	1 2

Key: 5 | 8 means 58

_____
_____

LESSON 7-1 Reteach
Frequency Tables, Stem-and-Leaf Plots, and Line Plots

A **frequency table** organizes data into groups.

A **cumulative frequency** table includes a running total of the frequencies of all the previous groups.

The list shows the pulse rate of 12 students.
    65, 63, 72, 88, 90, 82, 80, 75, 68, 72, 74, 84

You can use the data to make a cumulative frequency table. Here's how.

Step 1: Choose an interval. Each interval must be the same size. Often, 5 or 10 is a good interval length. Write the intervals in the first column.

Step 2: Find the number of values in each interval. Write the number in the middle column.

Step 3: Add the values to find the cumulative frequency. Write the number in the last column.

Pulse Rates

Pulse Rate	Frequency	Cumulative Frequency
60–69	3	3
70–79	4	4 + 3 = 7
80–89	4	4 + 7 = 11
90–99	1	1 + 11 = 12

intervals of 10 (pointing to Pulse Rate column)

the number of students with pulse rates between 80 and 89

the sum of the frequency column is equal to the last number in the cumulative frequency column.

1. The list below shows the highest temperatures (°F) ever recorded in all the states that begin with the letters M or N in the United States. Use the data to complete the cumulative frequency table.
105°, 109°, 107°, 112°, 114°, 115°, 118°, 117°, 118°, 125°, 106°, 110°, 122°, 108°, 110°, 121°

Highest Temperatures in M and N States

Temperature (°F)	Frequency	Cumulative Frequency
105–109	5	5
125–129		

Reteach
7-1 Frequency Tables, Stem-and-Leaf Plots, and Line Plots (continued)

A **stem-and-leaf plot** shows how often data values occur and how they are distributed. Numbers are separated into two parts: the stem and the leaves. The stem shows the leftmost digit in each number, and the leaves show the rightmost digit in each number.

You can use the pulse rate data to make a stem-and-leaf plot. Here's how.

65, 63, 72, 88, 90, 82, 80, 75, 68, 72, 74, 84

Step 1: List the stems in the left column of the plot.

Step 2: List the leaves in order in each row that corresponds to the stem of the number.

Step 3: Write a key and a title.

Pulse Rates

Stems	Leaves
6	3 5 8
7	2 2 4 5
8	0 2 4 8
9	0

Key: 7 | 2 means 72

The stems are the left digits of each number.

The leaves are the right digits of each number.

The first stem (6) has three leaves (3, 5, and 8). This represents the data values 63, 65, and 68.

---

2. The list below shows the number of stories of some of the tallest buildings in Houston, Texas. Use the data to complete the stem-and-leaf plot.

55, 52, 53, 64, 56, 71, 50, 46, 49, 75
50, 49, 46, 36, 47, 45, 47, 36, 44, 42

Tallest Buildings in Houston

Stems	Leaves
3	6
4	

Key:

Name _____ Date _____ Class _____

LESSON 7-1 Challenge
Read From the Middle Out

In a double stem-and-leaf plot, the stem is in the middle and the leaves are on both sides. You read from the middle to the left for the left data and the middle to the right for the right data.

The double stem-and-leaf plot compares the average monthly temperatures in Bloomington, Indiana, and Richmond, Virginia, in degrees Fahrenheit.

Average Monthly Temperatures (°F)

Bloomington		Richmond
Leaves	Stems	Leaves
7	2	
3 1	3	8
5 2	4	0 0 8 9
6 3	5	7 9
7 3	6	6
6 4 2	7	0 4 7 8

Key: 7 | 2 | means 27°F
Key: | 3 | 8 means 38°F

Use the double stem-and-leaf plot above to answer the questions.

1. What is the greatest average monthly temperature in Bloomington? in Richmond?

   _____

2. Which city had the lowest monthly temperature?

   _____

3. Which city has more months with monthly temperatures below 30°F?

   _____

4. Which city has more variation in temperatures? Explain.

   _____

5. What average temperature did Richmond have twice?

   _____

6. How many months was the average monthly temperature above 68° in Richmond?

   _____

7. What was the lowest average monthly temperature in Bloomington?

   _____

8. Compare the highest average monthly temperature in Bloomington to the highest average monthly temperature in Richmond.

   _____

Name _____ Date _____ Class _____

LESSON 7-1 Problem Solving
Frequency Tables, Stem-and-Leaf Plots, and Line Plots

Write the correct answer.

The table shows the time in minutes that Naima talked on the phone during the last 3 weeks.

Phone Time (min)

	Mon	Tues	Wed	Thurs	Fri	Sat	Sun
Week 1	12	15	25	45	52	30	31
Week 2	22	25	46	51	10	19	33
Week 3	44	21	30	20	10	24	52

1. Naima made a cumulative frequency table of the data using equal intervals. What number would she write in the frequency column for the interval 11–20 minutes?

2. Naima made a line plot of the data. Which numbers had more than one X above them?

3. If Naima makes a stem-and-leaf plot, which stem has the most leaves? What are they?

4. In the stem-and-leaf plot, which stems have the same number of leaves?

The list shows Hank Aaron's season home run totals. Make a cumulative frequency table, stem-and-leaf plot, and a line plot for the data. Then use the data to solve problems 5 – 8.

13, 27, 26, 44, 30, 39, 40, 34, 45, 44, 24, 32
44, 39, 29, 44, 38, 47, 34, 40, 20, 12, 10

5. In a cumulative frequency table of the data, what number belongs in the frequency column for interval 40-44?
   A 5
   B 6
   C 8
   D 14

6. In a cumulative frequency table of the data, what number belongs in the frequency column for interval 25-29?
   F 3
   G 5
   H 6
   J 8

7. In a stem-and-leaf plot of the data, how many stems do you need?
   A 1
   B 2
   C 3
   D 4

8. In a line plot of the data, which number would have 4 x's above it?
   F 34
   G 40
   H 44
   J 45

Name _____ Date _____ Class _____

LESSON 7-1 Reading Strategies
Use Graphic Aids

To the right is a listing of points scored in each of the league's basketball games.

28 48 34 50 47 35 40 37 36
55 43 39 43 34 52

Creating a **frequency table** is one way to organize the scores. A frequency table organizes scores by how often they occur.

League Basketball Scores

Points Scored	Frequency
20–29	1
30–39	6
40–49	5
50–59	3

Answer these questions about the frequency table.

1. How are the scores organized?

   _____

The **frequency,** or how often the scores occur, is recorded in the second column.

2. In how many games were 50 or more points scored?

   _____

3. In how many games were between 30 and 39 points scored? _____

A **stem-and-leaf plot** is another way to organize the data. Since the scores are two-digit numbers, the **stem** represents the tens digits, and the **leaves** represent the ones in each score.

Data in this row: 50, 52, 55

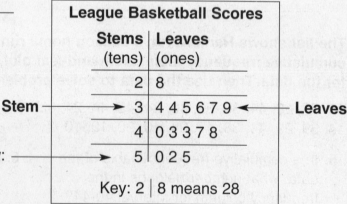

Answer each question about the stem-and-leaf plot.

4. Are the stems on the right or the left in a stem-and-leaf plot? _____

5. How many stems are in this plot? _____

6. What number does 3|6 represent? _____

10 Holt Mathematics

Puzzles, Twisters & Teasers
LESSON 7-1 *The Plot Thickens!*

Have you ever wondered about traveling trees? Finish this page to discover the answers to two important traveling tree questions! The list shows the number of minutes people spend watching television each day. Fill in the blanks to make a cumulative frequency table of the data.

10, 10, 15, 18, 25, 28, 32, 35, 35, 45, 50, 55, 57, 59

Minutes	Frequency	Cumulative Frequency
10–20		A
21–30		E
31–40		L
41–50		S
51–60		V

What does a tree do when it wants to go away?

It __L__ __E__ __A__ __V__ __E__ __S__ .
   9   6   4   14   6   11

Now use the list to make a stem-and-leaf plot.

Stems	Leaves
1	0 0 5(K) 8
2	5 8(N)
3	2(R) 5 5
4	5(T)
5	0 5 7(U) 9

Where does a traveling tree pack its things?

In its __T__ __R__ __U__ __N__ __K__ .
   45   32   57   28   15

Name _____ Date _____ Class _____

LESSON 7-2 Practice A
Mean, Median, Mode, and Range

Find the mean, median, mode, and range.

1. 5, 8, 5, 9, 3

2. 58, 54, 60, 56, 52

_____

3. 18, 17, 21, 18, 26

4. 60, 20, 40, 10, 50, 30

_____

The line plot below shows the number of kilometers Clara ran each day for 14 days. Use the line plot for Exercises 5 and 6.

5. Find the mean, median and mode for the set of data.
_____

6. Which measure of central tendency best describes the data? Explain your answer.
_____
_____

Use the data set to answer the questions.

22, 18, 15, 19, 61, 21

7. What is the outlier? _____

8. How does the outlier affect the mean, median, and mode?
_____
_____

9. Which measure of central tendency best describes the data set with the outlier?
_____

10. Which measure of central tendency best describes the data set without the outlier?
_____

Holt Mathematics

Name _____ Date _____ Class _____

Practice B
LESSON 7-2 Mean, Median, Mode, and Range

Find the mean, median, mode, and range of each data set.

1. 46, 35, 23, 37, 29, 53, 43

2. 72, 56, 47, 69, 75, 48, 56, 57

3. 19, 11, 80, 19, 27, 19, 10, 25, 15

4. 7, 8, 20, 6, 9, 11, 10, 8, 9, 8

5. The line plot shows the number of hours 15 students said they spent on homework in one week. Which measure of central tendency best describes the data? Justify your answer.

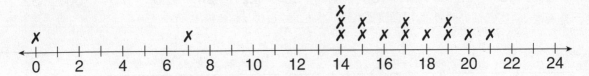

Identify the outlier in each data set. Then determine how the outlier affects the mean, median, and mode of the data. Then tell which measure of central tendency best describes the data with and without the outlier.

6. 14, 16, 13, 15, 5, 16, 12

7. 48, 46, 52, 92, 57, 58, 52, 61, 56

Holt Mathematics

Name _____ Date _____ Class _____

LESSON 7-2 Practice C
Mean, Median, Mode, and Range

Find the mean, median, mode, and range of each data set.

1. 21, 14, 17, 16, 23, 17, 15, 13

2. 44, 56, 38, 55, 62, 56, 48, 62, 56

3. The line plot shows Omar's point totals for 20 basketball games. Which measure of central tendency best describes the data? Justify your answer.

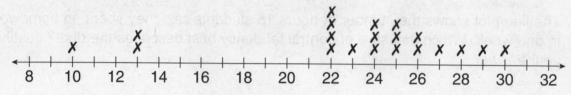

Identify the outlier in each data set. Then determine how the outlier affects the mean, median, and mode of the data. Then tell which measure of central tendency best describes the data with and without the outlier.

4. 26, 33, 36, 26, 30, 26, 10, 33

5. 59, 56, 62, 103, 61, 67, 62, 52, 63

6. Bart has these quiz scores: 78, 76, 74, 80, 82, 78, 82, 75, 77, 82. Bart tells his parents that his typical quiz score is an 82. Which measure of central tendency did Bart use to describe his typical score? Do you think Bart's measure of central tendency best described his typical score? Explain.

Name _____ Date _____ Class _____

LESSON 7-2 Reteach
Mean, Median, Mode, and Range

Measures of central tendency show what the middle of a data set looks like. The measures of central tendency are the *mean, median,* and *mode.*

Find the mean, median, mode, and range of 8, 3, 5, 4, 1, and 3.

Find the mean.
The mean is the sum of the values divided by the number of values in the data set.
1 + 3 + 3 + 4 + 5 + 8 = 24
24 ÷ 6 = 4
mean = 4

Find the range.
Find the difference between the least and greatest values.
8 − 1 = 7
range = 7

List in order: 1, 3, 3, 4, 5, 8

Find the mode.
The mode is the value that occurs most often. Sometimes there is no mode.
mode = 3

Find the median.
The median is the middle value.
median = 3.5

Find the range, mean, median, and mode of each data set.

1. 6, 5, 3, 6, 8

2. 12, 15, 17, 9, 17

3. 26, 35, 23, 27, 19, 23

4. 7, 6, 13, 16, 15, 9

5. 42, 38, 45, 42, 43

6. 51, 62, 68, 55, 68, 62

7. **Monthly Low Temperatures**

Month	Jun.	Jul.	Aug.	Sept.	Oct.
Temperature (°F)	44	41	47	42	36

Name _____ Date _____ Class _____

LESSON 7-2 Reteach
Mean, Median, Mode, and Range (continued)

An **outlier** is a value that is much greater than or much less than the other values in a data set.

How does the outlier affect the mean, median, and mode of the data?
  7, 9, 30, 9, 5, 6

Write the data in order and identify the outlier.
  5, 6, 7, 9, 9, **30** ← outlier

With the Outlier	Without the Outlier
Find the mean. 7 + 9 + 30 + 9 + 5 + 6 = 66 66 ÷ 6 = 11 The mean is 11.	Find the mean. 7 + 9 + 9 + 5 + 6 = 36 36 ÷ 5 = 7.2 The mean is 7.2.
Find the median. 5, 6, 7, 9, 9, 30 7 + 9 = 16 16 ÷ 2 = 8 The median is 8.	Find the median. 5, 6, 7, 9, 9 The median is 7.
The mode is 9.	The mode is 9.

8. How does the outlier affect the mean? _____

9. How does the outlier affect the median? _____

10. How does the outlier affect the mode? _____

To choose the measure of central tendency that best describes a set of data:

- Choose the measure that is closest to the greatest number of values in the data set, OR
- If there is an outlier, think about how it affects the mean and the median. Choose the measure that is affected least by the outlier.

Use the data set to answer the questions.
4, 6, 3, 6, 25, 3, 2

11. Is there an outlier? If so, what is it? _____

12. How does the outlier affect the mean and the median?

_____

13. Which measure of central tendency best describes the data? Explain your answer.

_____

_____

Name _____ Date _____ Class _____

LESSON 7-2 Challenge
These Puzzles Are Mean

Solve each puzzle.

1. There are 6 whole numbers in a set of numbers. The least number is 8, and the greatest number is 14. The mean, the median, and the mode are 11. What are the numbers?

   _____

2. There are 7 whole numbers in a set of numbers. The least number is 10, and the greatest number is 20. The median is 16, and the mode is 12. The mean is 15. What are the numbers?

   _____

3. There are 8 whole numbers in a set of numbers. The greatest number is 17, and the range is 9. The median and the mean are 12, but 12 is not in the data set. The modes are 9 and 14. What are the numbers?

   _____

4. The mean of a data set of 6 numbers is 8. The mean of a different data set of 6 numbers is 20. What is the mean of the combined data sets? _____

5. Find the mean of 7 numbers if the mean of the first 4 numbers is 5 and the mean of the last 3 numbers is 12. What is the mean of the combined data sets? _____

6. The mean of a data set of 3 numbers is 12. The mean of a data set of 9 numbers is 40. What is the mean of the combined data sets? _____

7. Sasha needs an average of 30 points to move to the next level in her competition. Her scores in the first three events are 28, 35, and 30. What is the lowest score she can score in her last event to move to the next level of competition? _____

8. Lars has a score of 89 for each of his first 3 science quizzes. The score on his fourth quiz is 92. What score does he need on his fifth quiz to have an average of 90? _____

9. Jake has a 95 average in math after 4 quizzes. Then he got 0 on the next quiz after being absent. There are 2 more quizzes. What average grade does he need on these last 2 quizzes to keep his average at least 85?

   _____

Name _____ Date _____ Class _____

Problem Solving
LESSON 7-2 *Mean, Median, Mode, and Range*

Write the correct answer.

The table to the right shows the leading shot blockers in the WNBA during the 2003 season.

Player	Shots Blocked
Margo Dydek	100
Lauren Jackson	64
Lisa Leslie	63
Ruth Riley	58
Michelle Snow	62

1. What is the range of this set of data?

   _____

2. What are the mean, median, and mode of this set of data?

   _____

3. What is the outlier in this set of data?

   _____

4. How does the outlier affect the mean and the median?

   _____

5. Which measure of central tendency best describes the set of data with the outlier? Explain.

   _____

Choose the letter for the best answer.

In a 100-meter dash, the first 5 racers finished with the following times: 11.6 seconds, 13.4 seconds, 10.8 seconds, 11.8 seconds, and 13.4 seconds.

6. Which measure of central tendency for this set of data is 12.2 seconds?
   A mean
   B median
   C mode
   D none of the above

7. Which measure of central tendency for this set of data is 11.8 seconds?
   F mean
   G median
   H mode
   J none of the above

8. What is the mode for this set of data?
   A 10.8 seconds
   B 11.8 seconds
   C 13.4 seconds
   D none of the above

9. The sixth racer finished with a time of 16.4 seconds. How will that affect the mean for this set of data?
   F decrease it by 0.7 second
   G increase it by 0.7 second
   H increase it by 3.28 seconds
   J does not affect the mean

Name _____ Date _____ Class _____

LESSON 7-2 Reading Strategies
Understanding Vocabulary

There are different ways to look at data. The **mean**, the **median**, the **mode**, and the **range** help us understand the information we gather.

The mean, or **average**, is the sum of the values divided by the total number of values.   Example: Ages in years → 18 11 6 12 18

   The average age is (18 + 11 + 6 + 12 + 18) ÷ 5 = 13 years

The **median** is the middle value in a set of data, when the values are listed in order. The median is 12.

   6   11   **12**   18   18

The **mode** is the most frequent value in a set of data. The mode is 18.

   6   11   12   **18   18**

The **range** indicates the spread of the data. It is the difference between the least and greatest values in a set of data.

   18 − 6 = **12**. The range is 12.

These are Jack's test scores for 5 math tests: 83, 79, 92, 95, 83. Use this data to help Jack look at his scores. Write "mean," "median," "mode" or "range" to answer the following questions.

1. 83 occurs more than any of the other values. Which measure is 83?

   _____

2. Jack added his test scores and divided by 5 to find which measure?

   _____

3. Jack found that the difference between his highest and lowest score was 16 points. What is this measure called?

   _____

4. Jack found that the middle value is 83. What is that value called?

   _____

Holt Mathematics

Name _____ Date _____ Class _____

LESSON 7-2 Puzzles, Twisters & Teasers
Stuck in the Middle!

Ever wonder what number in a data set sticks out like a sore thumb? To find out, use the data sets below to answer each question. Fill in the letter corresponding to each question above the question's answer in the decoder below.

1, 4, 2, 9, 5, 3

T: What is the range? _____

L: What is the mean? _____

I: What is the median? _____

6, 7, 4, 8, 12, 2, 1, 2

U: What is the range? _____

R: What is the mean? _____

O: What is the median? _____

E: What is the mode? _____

What number in some data sets sticks out like a sore thumb?

___ ___ ___ ___ ___ ___ ___
 5   11   8   4   3.5   2   5.25

Name _____ Date _____ Class _____

LESSON 7-3 Practice A
Bar Graphs and Histograms

The bar graph shows the lengths of four rivers. Use the graph for Exercises 1–3.

1. Which river is the longest?

   _____

2. About how much longer is the Amazon River than the Congo River?

   _____

3. About how much longer is the Nile River than the Huang River?

   _____

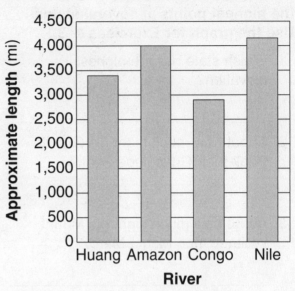

4. The table shows the ages of four U.S. Presidents when they first entered office. Make a bar graph of the data.

Name	President's Age
Truman	60
Kennedy	43
Carter	52
Bush	54

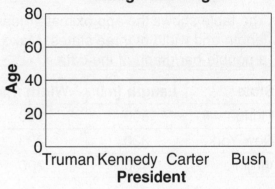

5. The table below shows the scores on a Spanish test. Make a histogram of the data.

Scores	Frequency
71–80	4
81–90	8
91–100	5

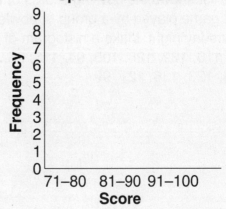

Name _____ Date _____ Class _____

LESSON 7-3 Practice B
Bar Graphs and Histograms

The bar graph shows the elevations of the highest points in several states. Use the graph for Exercises 1–3.

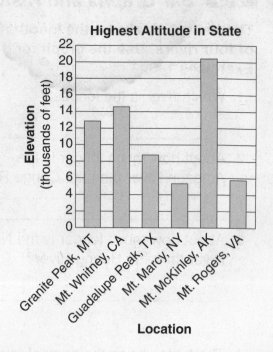

1. Which state has the highest elevation?

   _____

2. About how much higher is Granite Peak than Guadalupe Peak?

   _____

3. About how much higher is Mount Whitney than Mount Marcy?

   _____

4. The table shows the approximate mean length and width of three states. Make a double-bar graph of the data.

State	Length (mi)	Width (mi)
Florida	500	160
New York	330	283
Virginia	430	200

5. The list shows the bowling scores of the first game played by a group of bowlers on Thursday night. Make a histogram of the data.
96, 110, 132, 128, 105, 94, 116, 95, 126, 114, 123, 136, 121, 99

Name _____ Date _____ Class _____

LESSON 7-3 Practice C
Bar Graphs and Histograms

The bar graph shows the activities in which most injuries occur for children ages 5–14. Use the graph for Exercises 1–3.

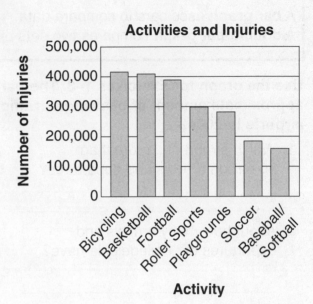

1. Which activity has the greatest number of injuries?

   _____

2. About how many more children are injured playing basketball than soccer?

   _____

3. About how many more children are injured participating in roller sports than on playgrounds?

   _____

4. The table shows the life expectancies of people living in three North American countries. Make a double-bar graph of the data.

Country	Male	Female
Canada	76	83
Greenland	66	74
Mexico	69	75

5. The list shows the times, in minutes, in which runners finished a 10-kilometer race. Make a histogram of the data.

   92, 97, 100, 88, 85, 79, 76, 83, 92, 87, 78, 85, 79, 98, 83, 84, 86

Name _____ Date _____ Class _____

LESSON 7-3 Reteach
Bar Graphs and Histograms

A **bar graph** uses bars to compare data.
A **double-bar graph** compares two sets of data.

Use the graph for Exercises 1–3. The bar graph shows the approximate amount of passenger traffic through some U. S. airports in 2004.

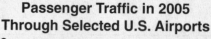

1. Which airport had more than 50 million arrivals and departures?

   _____

2. About how many arrivals and departures did Philadelphia have?

   _____

3. About how many more people passed through the Denver airport than the Orlando airport?

   _____

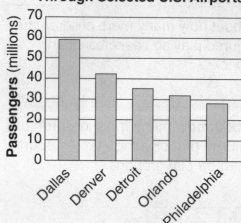

4. The table shows which sports are played by a seventh-grade class. Use the data in the table to make a double-bar graph.

Sport	Number of Boys	Number of Girls
Soccer	14	8
Tennis	7	10
Basketball	18	6
Swimming	12	20

   a. Finish labeling the vertical axis.

   What is the vertical axis label?

   _____

   What is the horizontal axis label?

   _____

   b. Draw a bar on the horizontal axis for the number of boys playing each sport. Draw a bar for the number of girls playing each sport. Label the bars.

   Each interval must be the same size.

   c. Give the graph a title. Make a key.

   Key: ☐

Holt Mathematics

Name _____ Date _____ Class _____

LESSON 7-3 Reteach
Bar Graphs and Histograms (continued)

A **histogram** is used to represent data in a frequency table. In a histogram the bars touch and both axes show numerical data.

Use the histogram for Exercises 5–7.

5. How many patients had a pulse rate between 80 and 89?

   _____

6. What range of pulse rates was most common among the patients?

   _____

7. How many more patients had a pulse rate between 100 and 109 than between 60 and 69?

   _____

The histogram shows the pulse rates of 20 patients during check-ups.

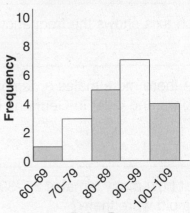

8. The frequency table shows the length (in inches) of the fish caught on a class field trip to the lake. Use the data in the frequency table to make a histogram.

   Fish Caught on Class Trip

Length (in.)	Frequency
1–5	1
6–10	6
11–15	8
16–20	5
21–25	2

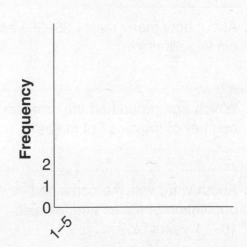

Fish Length (in.)

   a. Finish labeling the horizontal axis.

   What is the vertical axis label? _____

   What is the horizontal axis label? _____

   b. The vertical axis shows the frequency. Finish labeling the axis.

   c. Draw a bar for each interval. The bars should touch.

   d. Give the graph a title.

Holt Mathematics

Name _____ Date _____ Class _____

LESSON 7-3 Challenge
Sliding Histogram

The graph below is called a sliding histogram. It shows Germany's population in 2005.

Use the sliding histogram to the right to answer the questions.

1. What is the length of the interval used on the histogram?

   _____

2. Which axis shows the frequency?

   _____

3. Were there more males or females 85 years and older in Germany in 1997?

   _____

4. About how many females 35–39 years old were there?

   _____

5. About how many males 35–39 years old were there?

   _____

6. Which age group had the greatest number of females? of males?

   _____

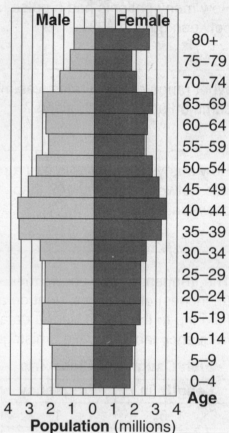

7. About what was the combined population of males and females 10–14 years old?

8. About how many more females 75–79 were there than males 75–79?

9. About how many more males were there 40–44 than 45–49?

10. At what age does the number of females start to differ significantly from the number of males?

Name _____ Date _____ Class _____

LESSON 7-3 Problem Solving
Bar Graphs and Histograms

Write the correct answer.

The double-bar graph shows the win-loss records for the Carolina Panthers football team in the years 1998-2004.

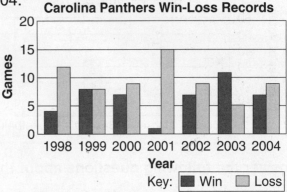

1. During how many seasons did the Panthers lose more games than they won?

2. In which year did the Panthers win more games than they lost?

3. Between which 2 years did the Panthers have the greatest improvement in their win-loss record?

4. In which year do you find the greatest range in the win-loss record?

Choose the letter for the best answer.

The histogram shows the ages of all members in a fan club.

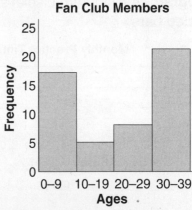

5. How many fan club members could be teenagers?
   A 5
   B 8
   C 17
   D 21

6. How many fan club members are between the ages of 30 and 39?
   F 5
   G 8
   H 17
   J 22

7. In which situation would you use a histogram to display data?
   A to show how you spend money
   B to show the change in temperature throughout the day
   C to show the golf scores from the whole team
   D to show the life expectancy of different animals

8. In which situation would you use a bar graph to display data?
   F to compare the speed of different computers
   G to show how a cat spends its time
   H to show how a child's height changes as he or she grows
   J to show the distribution of math grades in your class

Name _____ Date _____ Class _____

LESSON 7-3 Reading Strategies
Reading a Graph

Bar graphs create pictures for data.
A random sample was taken of boys' favorite sports.

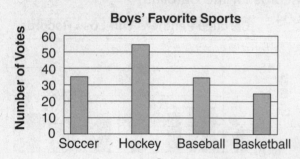

Answer the following questions about the bar graph.

1. What does the graph show? _____
2. What number does the scale of the graph count by?
   _____
3. How many boys like basketball best? _____
4. How many boys like hockey best? _____

Coaches were surveyed about the number of hours their teams practice each month. The data is shown in a frequency table and in a **histogram**. A histogram displays data in equal intervals with connected bars.

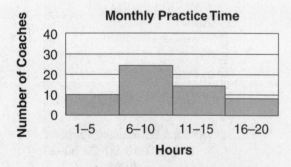

Answer the following questions about the histogram.

5. Where is the number of coaches located on the histogram?
   _____

6. The numbers on the left side of the histogram are in equal groups of what number?
   _____

7. What information is located along the bottom of the graph?
   _____

Name _____ Date _____ Class _____

LESSON 7-3 Puzzles, Twisters & Teasers
A Graphic Display!

Answer the question by studying the bar graph. Extract the letters from the graph to solve the riddle.

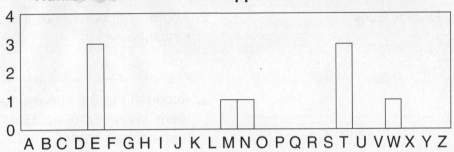

E = 1   M = 2   N = 3   T = 4   W = 5

What do you give a sick bird?

__ __ __ __ __ __ __ __ __
4 5 1 1 4 2 1 3 4

Name _____ Date _____ Class _____

LESSON 7-4 Practice A
Reading and Interpreting Circle Graphs

The circle graph directly below shows the results of a survey of 50 teens. They were asked about their favorite fruits. Use the graph for Exercises 1–3.

Favorite Fruits

[Circle graph showing Apples, Oranges, Grapes, Strawberries]

1. Did more teens pick apples or grapes?

   _____

2. About what percent of teens picked strawberries?

   _____

3. According to the survey, 10% of teens chose oranges. How many teens chose oranges?

   _____

The circle graph below shows the results of a survey of 100 people. They were asked about their favorite doughnut flavors. Use the graph for Exercises 4–6.

Favorite Doughnut Flavors

4. Did more people pick frosted or filled?

   _____

5. About what percent of people picked glazed?

   _____

6. According to the survey, 30% of the people chose filled doughnuts. How many people chose filled?

   _____

Would you use a bar graph or a circle graph to show the information? Explain your answer.

7. the number of bears that live in each of the national parks in Alaska

   _____
   _____
   _____

8. the number of visitors to Yellowstone National Park in August compared with the total number of summer visitors

   _____
   _____
   _____

Name _____ Date _____ Class _____

LESSON 7-4 Practice B
Reading and Interpreting Circle Graphs

The circle graph directly below shows the results of a survey of 80 teens who were asked about their favorite musical instruments. Use the graph for Exercises 1–3.

Favorite Musical Instruments

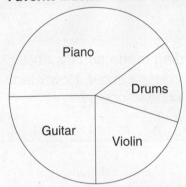

1. Did more teens pick piano or drums?

   _____

2. About what percent of teens picked guitar?

   _____

3. According to the survey, 20% of teens chose violin. How many teens chose violin?

   _____

The circle graph below shows the results of a survey of 100 people who were asked about their favorite vacation destinations. Use the graph for Exercises 4–6.

Favorite Vacation Destinations

4. Did more people pick mountains or beaches?

   _____

5. About what percent of people picked mountains?

   _____

6. According to the survey, 15% of the people chose famous landmarks. How many people chose famous landmarks?

   _____

Decide whether a bar graph or a circle graph would best display the information. Explain your answer.

7. number of tornadoes in each state during one year

   _____
   _____
   _____

8. the number of pounds of Macintosh apples sold compared with the total number of pounds of apples sold at a market in one day

   _____
   _____
   _____

Holt Mathematics

Name _____ Date _____ Class _____

LESSON 7-4 Practice C
Reading and Interpreting Circle Graphs

The circle graph directly below shows the results of a survey of 60 teens who were asked about their favorite sports. Use the graph for Exercises 1–3.

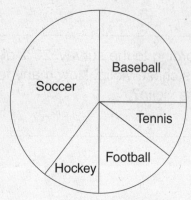
Favorite Sports

1. Did more teens pick baseball or hockey?

   _____

2. About what percent of teens picked tennis?

   _____

3. According to the survey, 40% of teens chose soccer. How many teens chose soccer?

   _____

The circle graph below shows the results of a survey of 200 people who were asked about their favorite breakfast foods. Use the graph for Exercises 4–6.

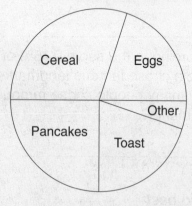
Favorite Breakfast Foods

4. Did more people pick cereal or eggs?

   _____

5. About what percent of people picked toast?

   _____

6. According to the survey, 25% of the people chose pancakes. How many people chose pancakes?

   _____

Decide whether a bar graph or a circle graph would best display the information. Explain your answer.

7. the amount of money budgeted for entertainment in one month compared with the total monthly allowance

   _____
   _____
   _____

8. the number of calories different birds of prey need to consume in one day

   _____
   _____
   _____

Name _____ Date _____ Class _____

LESSON 7-4 Reteach
Reading and Interpreting Circle Graphs

A **circle graph** shows how the parts of a complete set of data are related. The circle shows 100% of the data.

Choose a circle graph to show what percent (or part) of the total sales is represented by each type of pizza.

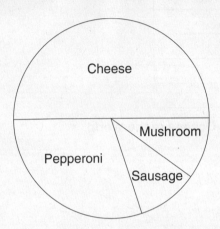

Choose a bar graph to show the number of each type of pizza sold.

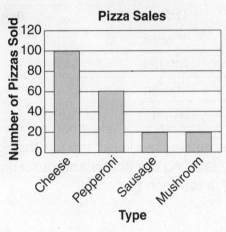

The circle graph shows that about one-half, or 50%, of the total sales was cheese pizzas. If 200 pizzas were sold, then about half, or 100, were cheese pizzas.

The circle graph shows the results of a survey of 100 teens who were asked about their favorite winter sport. Use the graph to answer each question.

Favorite Winter Sport

1. Did more teens pick skiing or snowboarding? _____

2. About what percent of teens picked skiing? _____

3. How many teens chose skiing? _____

Decide whether a bar graph or a circle graph would best display the information. Explain your answer.

4. the number of hurricanes in each state during 1 year

5. the number of students who play soccer compared with the total number of students in a school

_____   _____

_____   _____

_____   _____

Copyright © by Holt, Rinehart and Winston.
All rights reserved.

Holt Mathematics

Name _____ Date _____ Class _____

LESSON 7-4

Challenge
Circle the Oceans

The table below shows the approximate percent of area that each ocean comprises of the total area of the oceans of the world. These oceans together actually make up one large, connected body of water.

Ocean	Approximate Area
Pacific	49%
Atlantic	25%
Indian	22%
Arctic	4%

1. Complete the circle graph at the right by labeling the ocean and percent of area for each part.

Oceans of the World

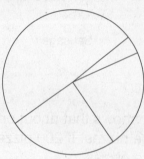

2. The oceans of the world cover almost 130 million square miles. About how many square miles are covered by the Pacific Ocean?

_____

3. About how many square miles are covered by the Atlantic Ocean?

_____

4. About how many square miles are covered by the Indian Ocean?

_____

5. About how many square miles are covered by the Arctic Ocean?

_____

6. About how many square miles are covered by the Pacific Ocean and the Atlantic Ocean combined?

_____

7. About how many more total square miles are covered by the Atlantic, Indian, and Arctic oceans than the by Pacific Ocean?

_____

Name _____ Date _____ Class _____

LESSON 7-4 Problem Solving
Reading and Interpreting Circle Graphs

Write the correct answer.

1. A market research group conducted a survey of 100 sports car owners. The group learned that 50% of the car owners loved their cars. What part of the circle in a circle graph would be represented by that statistic?

   _____

2. Juanita has 100 CDs. In her collection, 37 of the CDs are rock music, 25 are jazz, and 38 are country music. What part of the circle in a circle graph would represent the jazz CDs?

   _____

3. Mr. Martin wanted to compare his monthly rent to his total income. Should he use a circle graph or a bar graph?

   _____

4. Mr. Martin's rent has increased every year for the last 6 years. Should he use a circle graph or bar graph to show the yearly increase?

   _____

Choose the letter for the best answer. Use the circle graph.

5. To which age group do most of the fitness club members belong?

   A 18–20   C 30–39
   B 70+   D 40–49

Age of Fitness Club Members

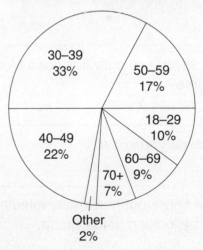

6. There are 100 members in a fitness club. How many members does the graph suggest will be between the ages of 18 and 39?

   F 10   H 43
   G 33   J 22

7. Which 2 age groups make up more than one-half the members?

   A 18–29 and 30–39
   B 30–39 and 40–49
   C 40–49 and 50–59
   D 18–29 and 70+

8. Which 2 age groups make up 3 times as many members as those who are between 60 and 69?

   F 40–49 and 50–59
   G 50–59 and 70+
   H 30–39 and 18–29
   J 18–29 and 50–59

35 Holt Mathematics

Name _____ Date _____ Class _____

LESSON 7-4
Reading Strategies
Reading a Graph

This **circle graph** gives you the whole picture of Molly's 60-minute exercise program. You can see how each part of her exercise program relates to the whole program.

Molly's 60-Minute Exercise Program

(Circle graph showing: Aerobics — half; Toning Exercises — one quarter; Stretching Exercises — one quarter)

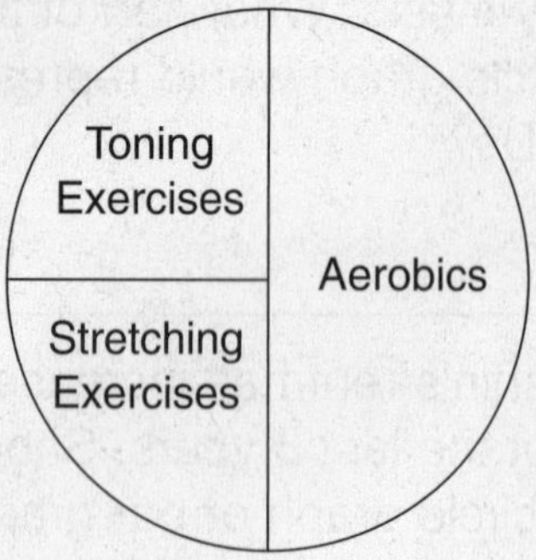

Answer the questions about the circle graph.

1. What does the graph show?
   _____

2. On which activity does Molly spend half of her time?
   _____

3. How do you know?
   _____

4. On which two activities does Molly spend the same amount of time?
   _____

5. How do you know?
   _____

You can figure out how many minutes Molly spends doing each activity by looking at the graph.

6. What is the total number of minutes Molly spends doing her full exercise program? _____

7. How many minutes does Molly spend doing aerobics? (She spends one-half of her exercise program doing aerobics.) _____

8. How many minutes does Molly spend doing stretching exercises? (She spends one-fourth of her time stretching.) _____

Name _____ Date _____ Class _____

LESSON 7-4 Puzzles, Twisters & Teasers
Cow-abunga!

Doctor Digby did a study of people's eating habits. He asked 100 people what kinds of meat they prefer. Fifty people said they like all kinds of meat. Twenty-five people said they don't like any kind of meat. Five people said they prefer fish. Ten people said they prefer chicken. Ten people said they prefer beef. Show the results of Doctor Digby's study as a circle graph using the empty circle below.

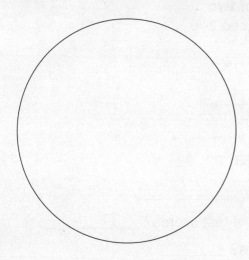

Use the key to answer the riddle: 5% = E 10% = D 25% = R 50% = U

Why did the cow cross the road?

To get to the __U__ __D__ __D__ __E__ __R__ side.
               50%   10%   10%   5%   25%

Name _____ Date _____ Class _____

LESSON 7-5 Practice A
Box-and-Whisker Plots

1. Use the data to make a box-and-whisker plot.
   24, 32, 35, 18, 20, 36, 12

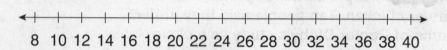

The box-and-whisker plot shows the test scores of two students. Use the box-and whisker plot for Exercises 2-4.

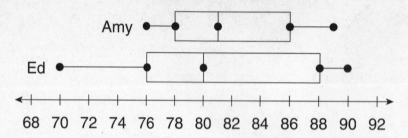

2. Which student has the greater median test score? _____

3. Which student has the greater interquartile range of test scores? _____

4. Which student has the greater range of test scores? _____

5. Which student appears to have more predictable test scores? Explain your answer.
   _____
   _____

The box-and-whisker plot shows prices of hotel rooms in two beach towns. Use the box-and whisker plot for Exercises 6-8.

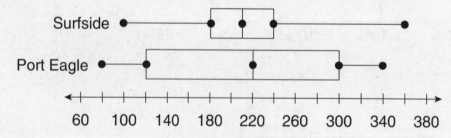

6. Which town has the greater median room price? _____

7. Which town has the greater interquartile range of room prices? _____

8. Which town appears to have more predictable room prices? Explain your answer.
   _____
   _____

Copyright © by Holt, Rinehart and Winston.
All rights reserved.

Holt Mathematics

Name _____ Date _____ Class _____

Practice B
7-5 Box-and-Whisker Plots

1. Use the data to make a box-and-whisker plot.
   19, 46, 37, 16, 24, 47, 23, 19, 31, 25, 42

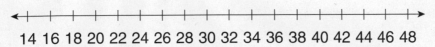

14 16 18 20 22 24 26 28 30 32 34 36 38 40 42 44 46 48

Use the box-and-whisker plot of games won per season by the New York Yankees and the Arizona Diamondbacks for 1998-2005 for Exercises 2-4.

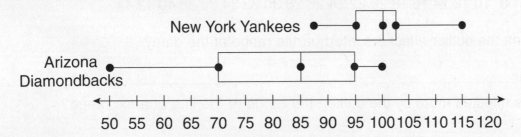

2. Which team has the greater median number of games won? _____
3. Which team has the greater interquartile range of games won? _____
4. Which team appears to have a more predictable performance? _____

Use the box-and-whisker plot of nightly tip totals that a waitress gets at two different restaurants for Exercises 5-7.

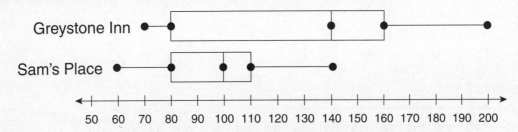

5. At which restaurant is the median tip total greater? _____
6. At which restaurant is the interquartile range of tip totals greater?
   _____
7. At which restaurant does the tip total appear to be more predictable?
   _____

Copyright © by Holt, Rinehart and Winston.
All rights reserved.

Holt Mathematics

Practice C
7-5 Box-and-Whisker Plots

Use the data for Exercises 1-3.
38, 42, 26, 32, 40, 28, 36, 27, 29, 6, 30

1. Make two box-and-whisker plots of the data on the same number line: one plot with the outlier and one plot without the outlier.

4 6 8 10 12 14 16 18 20 22 24 26 28 30 32 34 36 38 40 42 44

2. How does the outlier affect the interquartile range of the data?
   _____

3. Which is affected more by the outlier: the range or the interquartile range?
   _____

The table shows scores for two golfers. Use the table for Exercises 4-7.

| Henry | 78 | 80 | 74 | 91 | 73 | 88 | 92 | 94 | 78 | 80 |
| Trish | 82 | 84 | 81 | 82 | 80 | 89 | 86 | 90 | 78 | 85 |

4. Make two box-and-whisker plots of the data on the same number line.

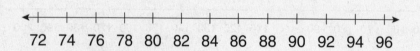

72 74 76 78 80 82 84 86 88 90 92 94 96

5. Which golfer has the lower median score? _____

6. Which golfer has the lesser interquartile range of scores? _____

7. Which golfer appears to be more consistent? _____

Name _____ Date _____ Class _____

LESSON 7-5 Reteach
Box-and-Whisker Plots

A **box-and-whisker plot** separates a set of data into four equal parts.

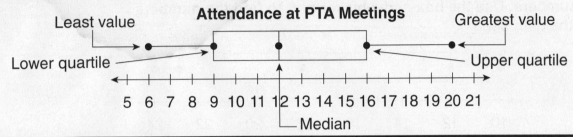

Use the data to create a box-and-whisker plot on the number line below:
35, 24, 25, 38, 31, 20, 27

1. Order the data from least to greatest.

2. Find the least value, the greatest value, and the median.

_____ _____

3. The **lower quartile** is the median of the lower half of the data. The **upper quartile** is the median of the upper half of the data. Find the lower quartile and the upper quartile in this data.

_____

4. Above the number line, plot points for the least value, lower quartile, median, upper quartile, and greatest value.

5. Draw a box around the quartiles and the median. Draw a whisker from the least value to the lower quartile. Draw a whisker from the upper quartile to the greatest value.

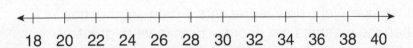

Use the data to create a box-and-whisker plot:
63, 69, 61, 74, 78, 72, 68, 70, 65

6. Order the data. _____

7. Find the least and greatest values. _____

8. Find the median, lower quartile, and upper quartile. _____

9. Plot the points, draw the box, and add the whiskers.

⟵—|—|—|—|—|—|—|—|—|—|—⟶
60 62 64 66 68 70 72 74 76 78 80

Copyright © by Holt, Rinehart and Winston.
All rights reserved.

Holt Mathematics

Name _____ Date _____ Class _____

LESSON 7-5 Challenge
Puzzling Plots

The box-and-whisker plot below represents a data set that has 7 numbers. Use the box-and-whisker plot to find the numbers in the data set.

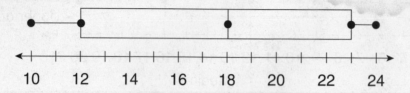

1. What are the least and greatest numbers in the data set? _____

2. What is the median? _____

3. Is the median one of the 7 numbers in the data set, or is it the mean of the middle two numbers. Explain your answer.

   _____
   _____

4. What are the lower and upper quartiles? _____

5. Are the lower and upper quartiles two of the numbers in the data set? Explain your answer.

   _____
   _____

6. The remaining two numbers in the data set have a mean of 19 and range of 6. What are the two numbers? _____

7. The box-and-whisker plot below represents a data set that has 7 numbers. When the numbers are in order from least to greatest, the third and fifth numbers in the set have a mean of 12 and a range of 4. What are the 7 numbers?

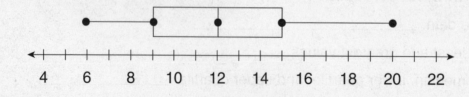

_____

Copyright © by Holt, Rinehart and Winston.
All rights reserved.

Holt Mathematics

Name _____ Date _____ Class _____

LESSON 7-5 Problem Solving
Box-and-Whisker Plots

Write the correct answer.

A fitness center offers two different yoga classes. The attendance for each class for 12 sessions is represented in the box-and-whisker plot.

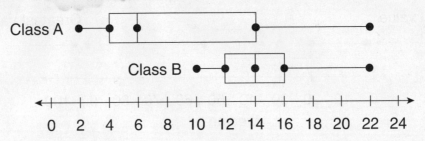

1. Which class has a greater median attendance? How much greater is it?

2. Which class appears to have a more predictable attendance?

3. Which class has an attendance of less than 14 people 75% of the time?

4. What percent of the time does Class B have an attendance greater than 16?

Choose the letter for the best answer.

The box-and-whisker plot shows the percent of people in eight Central American countries who used the Internet in 2005.

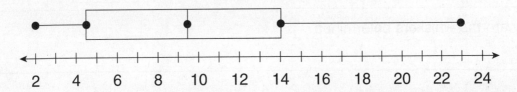

5. What is the range in the percents of people who used the Internet in the eight countries?
   A 23%
   B 21%
   C 14%
   D 9.5%

6. In how many of these eight countries did at least 9.5% of the people use the Internet?
   F less than 2
   G less than 4
   H at least 4
   J cannot be determined

7. What is the mode of the data?
   A 2%
   B 4.5
   C 9.5%
   D cannot be determined

8. What is the interquartile range of the data?
   F 23%
   G 21%
   H 14%
   J 9.5%

Name _____ Date _____ Class _____

LESSON 7-5 Reading Strategies
Use Graphic Aids

A **box-and-whisker plot** shows a set of data divided into four equal parts called **quartiles**.

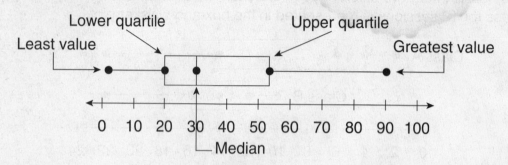

- The median score divides the set of data in half. The median score for this plot is 30.

- The box shows the middle half of the data, located on either side of the median. The box extends from 20 to 53.

- The two whiskers identify the remaining half of the data. One whisker extends from the box to the greatest value: from 53 to 90. The other whisker extends from the box to the least value: from 2 to 20.

Answer each question.

1. What does the box stand for in a box-and-whisker plot?

   _____

2. How are the whiskers determined?

   _____

   _____

3. Why is it important to find the median score?

   _____

Describe where these scores are located in the box-and-whisker plot above.

4. 18   between the _____ and the _____.

5. 75   between the _____ and the _____.

6. 45   between the _____ and the _____.

Name _____ Date _____ Class _____

Puzzles, Twisters & Teasers
LESSON 7-5 *The Plot Thickens!*

Circle words from the word list in the word search below.

box   whisker   plot   first   third   upper
quartile   extreme   data   median   range   fraction

```
E B O X Z D W P Y J I O R C
X F D Q O W E R F C O D A T
P R R U E H H T I P J A D A
L A D A W T H I R D U T B M
O C A R C S B I S R P A I E
T T O T A T A L T K P G H D
U I K I O N M T N F E A Z I
P O M L D E G R D G R R W A
L N B E X T R E M E E S V N
```

Now think about the alphabet and the letters you see above.

What two letters of the alphabet hold nothing when you say them together?

_____ _____

Name _____ Date _____ Class _____

LESSON 7-6
Practice A
Line Graphs

Use the table to answer the questions.

U.S. Personal Spending on Selected Electronics	
Year	Amount Spent ($billions, estimated)
1994	$71
1996	$80
1998	$90
2000	$107

1. Use the data in the table to make a line graph.

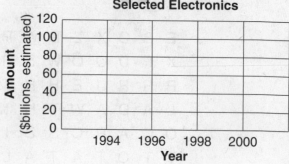

2. When did the amount spent on electronics increase the most?

_____

3. About how much was spent on electronics in 1999?

_____

The table below shows normal monthly temperatures in Asheville, North Carolina, and Miami, Florida, for the first 4 months of the year. Use the table to answer the questions.

	Jan.	Feb.	Mar.	Apr.
Asheville	36°F	39°F	47°F	55°F
Miami	67°F	69°F	72°F	75°F

4. Make a double-line graph of the data.

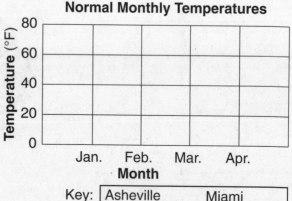

5. During which month is there the greatest difference in temperatures between the two cities?

_____

6. During which month is there the least difference in temperatures between the two cities?

_____

7. In which city did the temperature vary the least during these 4 months?

_____

8. Where should you spend your April vacation if you want to go hiking wearing your new sweatshirt?

_____

Copyright © by Holt, Rinehart and Winston.
All rights reserved.

Holt Mathematics

Name _____ Date _____ Class _____

LESSON 7-6 Practice B
Line Graphs

Use the table for Exercises 1–3.

Retail Price of Regular Gasoline in the United States (to the nearest cent)

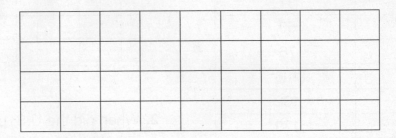

Year	1990	1992	1994	1996	1998	2000	2002	2004
Price Per Gallon	$1.16	$1.13	$1.11	$1.23	$1.06	$1.51	$1.32	$1.82

1. Make a line graph of the data.

2. When did the cost of gasoline decrease the most?

3. About how much did gasoline cost in 1995?

_____

The table below shows the student population at elementary schools in two cities, New City and Jackson.

Year	1996	1997	1998	1999	2000	2001	2002	2003
New City	450	460	440	430	495	500	600	645
Jackson	500	475	450	525	430	440	485	480

4. Make a double-line graph of the data.

5. During which year did New City's school population increase the most?

6. The mall in Jackson closed. Many people lost their jobs and moved their families to New City, where a new mall opened. In what year did this probably happen? Explain your thinking.

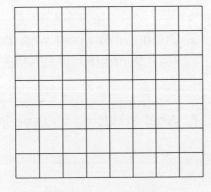

Key: _____

Name _____ Date _____ Class _____

LESSON 7-6 Practice C
Line Graphs

Use the table for Exercises 1–3.

| U.S. Population (to the nearest million) ||
Year	Number
1820	10
1840	17
1860	31
1880	50
1900	76
1920	106
1940	132
1960	179
1980	226
2000	281

1. Make a line graph of the data.

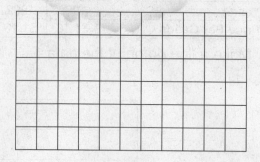

2. When did the U.S. population have the greatest increase?

_____

3. Estimate the U.S. population in 1970.

_____

The table below shows the price in cents per pound U.S. farmers received for apples and peanuts they grew.

Year	1996	1997	1998	1999	2000	2001	2002	2003
Apples (cents/lb)	20.8	22.1	17.1	21.3	17.8	22.9	25.6	29.5
Peanuts (cents/lb)	28.1	28.3	25.7	25.4	27.4	23.4	18.2	18.8

4. Make a double-line graph of the data in the table.

5. During which year was there the greatest difference in prices?

_____

6. During which year was there the least difference in prices?

_____

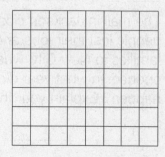

Key: [ ]

7. What trends do you see in the data?

_____
_____
_____

Name _____ Date _____ Class _____

LESSON 7-6 Reteach
Line Graphs

A **line graph** shows data changing over time.
A **double-line graph** shows two sets of data changing over time.

1. The table shows the normal daily low temperature in °F in Indianapolis and San Francisco. Use the data to make a double-line graph.

Month	Jan.	Mar.	May	Jul.	Sept.	Nov.
Indianapolis	26°	41°	63°	75°	67°	43°
San Francisco	49°	53°	58°	63°	65°	55°

a. Finish labeling each axis.

   What is the vertical axis label?

   _____

   What is the horizontal axis label?

   _____

b. Plot each point for each set of data.

c. Connect the points. Use a dotted line to represent one set of data.

d. Give the graph a title and key.

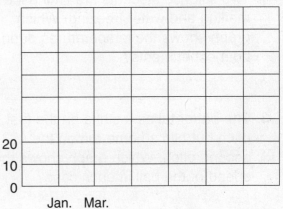

Key: _____

2. In which city was the temperature the most constant?

   _____

3. During which 2 months in Indianapolis did the temperature increase the most?

   _____

4. Use the graph to estimate the low temperature in October in San Francisco.

   _____

5. During which month was the difference in the daily normal low temperatures in Indianapolis and San Francisco most likely the smallest?

   _____

Holt Mathematics

Name _____ Date _____ Class _____

LESSON 7-6 Challenge
What's My Line?

Write the letter of the graph that best represents each event described.

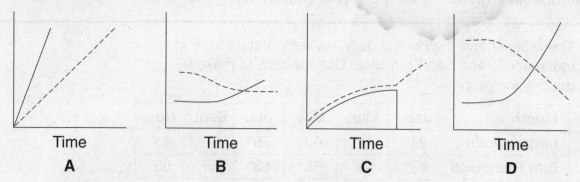

1. Two science students are timing ice melting and water freezing. Which graph shows the temperatures during both experiments?

2. In the music industry, when CDs became popular, sales of cassettes plummeted. Which graph shows this trend?

3. The batter takes a strike on the first pitch but hits a home run on the second pitch. Which graph shows the speed of the ball in each case?

4. Two brothers attend the same school. One walks and the other rides a bike. Which graph compares the time required for them to travel to school?

Draw a double-line graph to solve.

5. The town swimming pool puts aside the last 15 minutes of every hour for an adult swim. Show the change in the number of children and adults in the pool over several hours.

6. The Ice Cream Station is open only during the summer months. Cool Guy's Ice Cream is open all year long. Show the change in ice cream sales over the course of a year for each business.

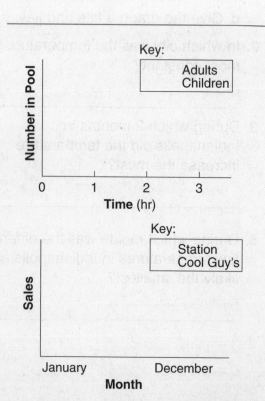

Name _____ Date _____ Class _____

LESSON 7-6 Problem Solving
Line Graphs

Write the correct answer.

The line graph shows the number of households with cable television from 1996 to 2002.

1. About how many households had cable TV in 1998?

   _____

2. About how many more households had cable TV in 2002 than in 1996?

   _____

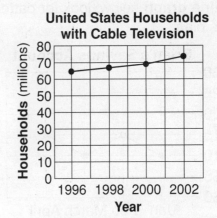

United States Households with Cable Television

3. During which two-year period did the number of households with cable TV grow the most?

   _____

4. Use the graph to estimate the number of households with cable TV in 2001.

   _____

Choose the letter for the best answer.

The double-line graph shows the number of tornadoes in the United States during part of 2002 and 2003.

5. About how many more tornadoes were there in August 2002 than in August 2003?

   A about 50   C about 30
   B about 40   D about 20

6. In which time period did the number of tornadoes increase both years?

   F Sept. to Oct.   H Nov. to Dec.
   G Oct. to Nov.   J none

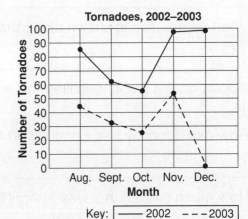

Tornadoes, 2002–2003

Key: —— 2002 - - - 2003

7. Which time period showed the greatest decrease in the number of tornadoes?

   A Aug. to Sept. 2002
   B Aug. to Sept. 2003
   C Oct. to Nov. 2002
   D Nov. to Dec. 2003

8. During which month was there the greatest difference between the number of tornadoes in 2002 and in 2003?

   F Aug.   H Nov.
   G Sept.   J Dec.

Copyright © by Holt, Rinehart and Winston.
All rights reserved.

Holt Mathematics

Name _____ Date _____ Class _____

LESSON 7-6 Reading Strategies
Organization Patterns

A **line graph** visually shows data that changes over time. Your height changes from year to year. Sales at a store change from month to month.

In the **line graph** below, look for patterns of change in a savings plan.

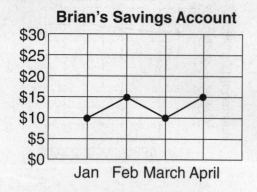

Answer the following questions about the line graph.

1. What information is located along the left side of the graph?

   _____

2. In what intervals are the numbers shown on the left side of the graph?

   _____

3. What information is shown along the bottom of the graph?

   _____

Each point on the graph identifies the amount of money saved each month.

4. How much money had Brian saved in February?

   _____

5. How much had Brian saved in April?

   _____

6. Why do you think Brian's savings account dropped between February and March?

   _____

Puzzles, Twisters & Teasers
LESSON 7-6 Rhyme Time TV!

Make a line graph of the data in the table. Think about the data and the way it is presented. Then make a rhyme to answer the riddle by filling in the letter associated with each ordered pair.

What do you call a table that shows information about cable TV?

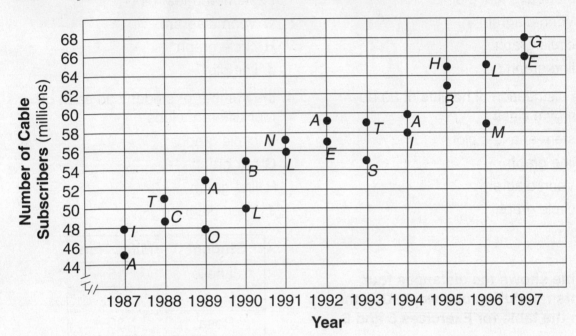

Year	Number of Cable Subscribers (millions)	Letter to Solve Riddle
1987	45	A
1988	49	C
1989	53	A
1990	55	B
1991	56	L
1992	57	E
1993	59	T
1994	60	A
1995	63	B
1996	65	L
1997	66	E

Answer: A CABLE TABLE

Name _____ Date _____ Class _____

LESSON 7-7 Practice A
Choosing an Appropriate Display

Choose the letter of the graph that would best represent each type of data.

1. the number of visitors to a national park in each of the past five years
   A stem-and-leaf plot
   B Venn diagram
   C circle graph
   D line graph

2. the part of a whole group that said yes, no, or no opinion in a survey
   F stem-and-leaf plot
   G Venn diagram
   H circle graph
   J line plot

3. the distribution of heights of 30 boys in a gym class
   A stem-and-leaf plot
   B line graph
   C Venn diagram
   D circle graph

4. the number of students in each of four different clubs
   F circle graph
   G bar graph
   H line graph
   J Venn diagram

The table shows the distances four students ran during one week. Use the data in the table for Exercises 5 and 6.

Student	Miles Run
Janelle	15
Marty	25
Rosa	30
Sidney	10

5. Does the stem-and leaf plot do a good job of representing the data from the table? Explain your answer.

   Distances Run

Stems	Leaves
1	0 5
2	5
3	0

   Key: 1|2 means 12 miles

   _____
   _____
   _____

6. Does the bar graph do a good job of representing the data? Explain.

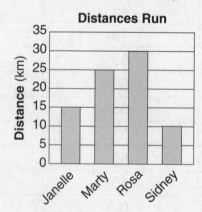

   _____
   _____
   _____

Copyright © by Holt, Rinehart and Winston.
All rights reserved.

Holt Mathematics

Name _____ Date _____ Class _____

Practice B
LESSON 7-7 Choosing an Appropriate Display

Choose the type of graph that would best represent this data.

1. the number of points scored by five different basketball players in a game

   _____

2. the distribution of test scores in a math class

   _____

3. the students who are in the chess club, the debating club, and the computer club, and the students who are in more than one of those clubs

   _____

4. the percent of total income a family uses for rent, food, clothing, entertainment, savings and other expenditures

   _____

The table shows the earnings of The Sandman Company during each of five years. Explain why each display does or does not appropriately represent the data.

Year	Earnings (millions of dollars)
2001	1
2002	3
2003	8
2004	12
2005	16

5. **Earnings of The Sandman Company**

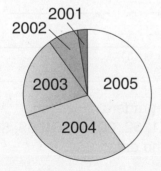

_____
_____
_____

6. **Earnings of The Sandman Company**

_____
_____
_____

Copyright © by Holt, Rinehart and Winston.
All rights reserved.

Holt Mathematics

Name _____ Date _____ Class _____

LESSON 7-7 Practice C
Choosing an Appropriate Display

Choose the type of graph that would best represent this data.

1. the distribution of scores of a professional golfer for 20 rounds of golf

2. the change in weight over time when the diet of a dog is changed.

_____

_____

The table shows the responses of 100 students to a survey. Explain why each display does or does not appropriately represent the data.

3. **Favorite Sport to Play**

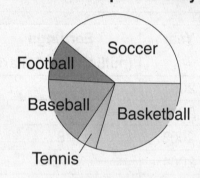

Favorite Sport to Play	Number of Students
Baseball	15
Basketball	30
Football	10
Soccer	40
Tennis	5

4.

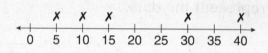

_____

_____

_____

_____

_____

_____

5. The data gives the heights in inches of 20 seventh-graders.
58, 64, 60, 56, 66, 61, 60, 58, 70, 62, 60, 58, 56, 71, 62, 60, 61, 64, 59, 56
Make the type of graph that would best represent the data.

Name _____ Date _____ Class _____

LESSON 7-7 Reteach
Choosing an Appropriate Display

There are different ways to show data. To choose a data display, think about the information that each kind of graph shows.

Bar graph: Compares amounts.

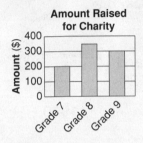

Circle graph: Shows data as parts of a whole.

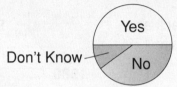

Stem-and-leaf plot: Shows each value and how the values are distributed.

Test Scores

Stems	Leaves
6	8
7	5 6 6 8 9 9
8	2 2 4 4 5 8 8
9	1 7

Key: 6|8 means 68

Line plot: Shows each value and how the values are distributed.

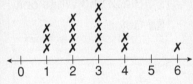

Line graph: Shows how amounts change over time.

Venn diagram: Shows members of sets and how sets overlap.

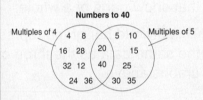

Circle the type of graph which would best show the data. Explain the reason for your choice.

1. the way the vote is split among three candidates in a school election
   circle graph or stem-and-leaf plot

   _____
   _____
   _____

2. the average income for a family of four in 1970, 1980, 1990, and 2000
   line plot or line graph

   _____
   _____
   _____

3. the points scored by a football team during each game in a 16-game season
   stem-and-leaf plot or circle graph

   _____
   _____
   _____

4. the calories per ounce in four different cuts of meat
   line plot or bar graph

   _____
   _____
   _____

Copyright © by Holt, Rinehart and Winston.
All rights reserved.

57

Holt Mathematics

LESSON 7-7

Challenge
Divided Bar Graphs

Suppose you wanted to show how parts of whole change over time. One way to do this would be to draw a series of circle graphs.

These circle graphs show the percent of TV households that have color TVs or only black and white TVs.

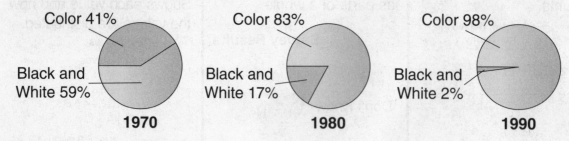

Another display you could use is a divided bar graph. In a divided bar graph, each bar is divided into sections that show parts of a whole.

This divided bar graph represents the same data as the three circle graphs above.

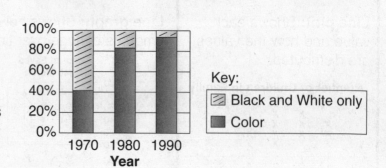

The table shows how many hours of television students at one high school watched. Use the data from the table to complete the divided bar graph.

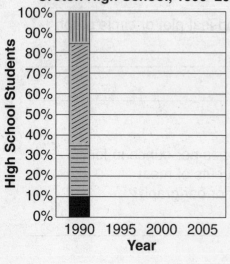

Television Viewing Habits at Croton High School, 1990-2005

	1990	1995	2000	2005
1 hour or less	40	44	70	120
1 to 2 hours	100	124	175	180
2 to 3 hours	200	176	80	50
More than 3 hours	60	56	50	30

Key:
- More than 3 hours
- 2–3 hours
- 1–2 hours
- 1 hour or less

Name _____ Date _____ Class _____

Problem Solving
LESSON 7-7 Choosing an Appropriate Display

Write the correct answer.

1. You take a survey of your class to find the number of years each student has lived in your town. You want to show how the data is distributed. What kind of graph would you use to display your data? Explain your choice.

   _____
   _____
   _____

2. Wendy budgets $120 for the week. $30 is for transportation, $50 is for food, $25 is for entertainment, and $15 is for other expenses. What kind of graph would best represent the data? Justify your response.

   _____
   _____
   _____

3. The New York Mets' worst season was their first season, in 1962. In that year they won 40 games. Their best season was in 1986, when they won 108 games. You want to show the number of games won each year in a way that makes it easy to see the distribution of the data. Why might you use a stem-and-leaf plot rather than a line plot?

   _____
   _____
   _____

4. During one school year, Matt reads 16 books, Tama reads 22 books, Rhonda reads 14 books, and Francisco reads 20 books. Would a line graph be an appropriate way to display this data? Explain your answer.

   _____
   _____
   _____

Choose the letter for the best answer.

5. Which type of graph would be most appropriate to show the distribution of daily high temperatures for a month?

   A circle graph
   B Venn diagram
   C stem-and-leaf plot
   D line graph

6. Which type of graph would be most appropriate to compare the price of the same TV at five discount stores?

   F stem-and-leaf plot
   G bar graph
   H circle graph
   J line plot

Name _____ Date _____ Class _____

LESSON 7-7 Reading Strategies
Use a Flowchart

You can use a flowchart to help you choose an appropriate display for a set of data.

Start Here

Do you need to show amounts that change over time? —Yes→ Use a line graph.

↓ No

Do you need to compare amounts? —Yes→ Use a bar graph.

↓ No

Do you need to show parts of a whole? —Yes→ Use a circle graph.

↓ No

Do you need to show two or more sets of data and how they overlap? —Yes→ Use a Venn diagram.

↓ No

Do you need to see how often each piece of data occurs? —Yes→ Use a line plot or a stem-and-leaf plot.

Use the flowchart and the situation below for Exercises 1 and 2.

For a TV show, 55% of the viewers are under 18, 35% are between 18 and 34, and 10% are over 34. You want to show how big a part of the whole audience each age group is.

1. For which question in the flowchart is the answer yes?

   _____

2. Which kind of graph will you use? _____

Use the flowchart and the situation below for Exercises 3 and 4.

A school has 34 different classes with sizes between 18 and 28 students. You want to show each of the different class sizes.

3. For which question in the flowchart is the answer yes?

   _____

4. Which kind of graph will you use? _____

Name _____ Date _____ Class _____

LESSON 7-7 Puzzles, Twisters, & Teasers
Seeing Is Believing!

Choose the type of graph that would be best for each situation.
Circle the letter above your answer.

1. comparing the numbers of raffle tickets sold by four students
   - **L** line plot
   - **O** bar graph

2. displaying every winning speed in the history of the Indianapolis 500 car race
   - **Y** stem-and-leaf plot
   - **D** circle graph

3. showing the factors and common factors of 24, 30, and 35
   - **A** circle graph
   - **G** Venn diagram

4. showing the sales of a monthly magazine in its first six months
   - **R** Venn diagram
   - **E** line graph

5. comparing the parts of a monthly budget
   - **D** circle graph
   - **T** stem-and-leaf plot

6. displaying quiz scores for 20 quizzes on which the scores can be from 1-10
   - **S** line plot
   - **P** line graph

Use the circled letters above the problem numbers to solve the riddle.

What do you need to spot an iceberg from 10 miles away?

__ __ __ __ __ __ __ __
3. 1. 1. 5.   4. 2. 4. 6.

Name _____ Date _____ Class _____

LESSON 7-8 Practice A
Populations and Samples

Choose the letter of the sampling method that will better represent the whole population.

1. Clinton School Cafeteria: Student Satisfaction

    a. Mark surveys 40 students who are in his classes. 72% are satisfied with the food in the cafeteria.

    b. Tammy surveys 65 students by randomly choosing names from a list of all students in the school. 85% are satisfied with the food in the cafeteria.

2. Predicted Winner in an Election for Mayor

    a. Harris telephones 100 randomly chosen voters. 54% plan to vote for Mayor Green.

    b. Julia asks 70 people whom she knows. 45% plan to vote for Mayor Green.

For Problems 3-5, tell whether each sample may be biased. Explain your answer.

3. A town official surveys 50 people in a library to decide if town residents want the library expanded.

   _____
   _____

4. A cable TV company randomly calls 200 customers and asks them if they are satisfied with their service.

   _____

5. George surveys 15 students on the soccer team to learn whether middle school students want more money spent on school sports.

   _____
   _____

6. A factory produces 12,000 computers per week. The manager of a factory claims that fewer than 50 defective computers are produced each week. In a random sample of 500 computers, 2 were defective. Tell if the manager's claim is likely to be true.

   _____

Name _____ Date _____ Class _____

LESSON 7-8 Practice B
Populations and Samples

1. Determine which sampling method will better represent the entire population. Justify your answer.

Reading Habits of High School Students	
Sampling Method	Results of Survey
Dinah surveys 48 students who she knows.	91% have read a novel in the past month.
Suki gives survey forms to 100 students who were randomly chosen from a school attendance list.	59% have read a novel in the past month.

_____
_____

For Problems 2 and 3, determine whether each sample may be biased. Explain.

2. An on-line bookseller randomly chooses 200 book buyers from its database and then surveys those book buyers to find out if they were satisfied with the time it took to deliver their orders.

_____
_____

3. Milena surveys 80 high school students who are leaving a jazz concert to determine the favorite type of music among high school students.

_____
_____

4. Zack chooses a random sample of 50 out of 400 students. He finds that 7 of them have traveled to a foreign country. Zack claims that over 50 of the 400 students have traveled to a foreign country. Do you agree? Explain your answer.

_____
_____

5. A mint produces 150,000 souvenir coins each year. In a random sample of 400 coins, 3 have a misprint. Predict the number of coins that will have misprints in a year.

_____

Name _____ Date _____ Class _____

LESSON 7-8 Practice C
Populations and Samples

1. Determine which sampling method will better represent the entire population. Justify your answer.

The Midland Company: Employee Satisfaction	
Sampling Method	Results of Survey
Wanda interviews 80 employees who were randomly selected from the company payroll list.	69% feel adequately challenged by their jobs.
Bernard interviews the last 30 employees to be hired.	90% feel adequately challenged by their jobs.

_____
_____
_____

For Problems 2 and 3, determine whether each sample may be biased. Explain.

2. A landlord e-mails 60 of his 1,250 tenants and surveys them to determine whether they would like to use the Internet to pay rent.

_____
_____

3. An insurance company surveys 350 of its customers by randomly choosing names from its customer database and then telephoning the customers.

_____

Explain whether you would survey the entire population or use a sample.

4. You want to know how many hours members of a sports team train each week during the off-season.

_____

5. You want to know the average income of people who eat at vegetarian restaurants across the country.

_____

LESSON 7-8 Reteach
Populations and Samples

Survey topic: number of books read by seventh-graders in Richmond

A **population** is the whole group that is being studied.	*Population*: all seventh-graders in Richmond
A **sample** is a part of the population.	*Sample*: all seventh graders at Jefferson Middle School
A **random sample** is a sample in which each member of the population has a random chance of being chosen. A random sample is a better representation of a population than a non-random sample.	*Random sample*: Have a computer select every tenth name from an alphabetical list of each seventh-grader in Richmond.
A **biased sample** is a sample that does not truly represent a population.	*Biased sample*: all of the seventh graders in Richmond who are enrolled in honors English classes.

Tell if each sample is biased. Explain your answer.

1. An airline surveys passengers from a flight that is on time to determine if passengers on all flights are satisfied.

   _____

   _____

2. A newspaper randomly chooses 100 names from its subscriber database and then surveys those subscribers to find if they read the restaurant reviews.

   _____

   _____

3. The manager of a bookstore sends a survey to 150 customers who were randomly selected from a customer list.

   _____

4. A team of researchers surveys 200 people at a multiplex movie theater to find out how much money state residents spend on entertainment.

   _____

   _____

LESSON 7-8

Challenge
A Sample of Samples

Consider the three sampling methods described below.

Systematic sample: You randomly select an individual and then follow a pattern to select others in the sample. For example, choose a name from the first 50 names in a telephone book. Then choose every fiftieth name after that.

Stratified sample: A population is divided into subgroups, called strata, that contain similar individuals. For example, interview all the boys and girls at a movie. Boys and girls are two strata.

Cluster sample: A population is divided into sections and then a few of the sections are selected. All the members of those sections are chosen. For example, suppose a city has 50 voting districts. Five of those districts are selected, and every person in each district selected is polled.

Identify the type of sampling used as systematic, stratified, or cluster.

1. A teacher selects every third student in the class. _____

2. A teacher surveys all students from each of 5 randomly selected classes. _____

3. A principal selects 3 girls and 3 boys from each of 10 classes. _____

4. A teacher selects 10 students under 12 years old and 10 students over 12 years old. _____

5. A reporter interviews all students in each of 4 randomly selected schools. _____

6. A reporter interviews 20 men and 20 women. _____

7. Workers on an assembly line check every tenth tire in a tire manufacturing plant. _____

8. An advertising consultant surveys all members in each of 3 randomly selected fitness clubs from all the fitness clubs in the city. _____

9. A department store manager selects a customer from the first 20 on a customer list. Then she selects every twentieth customer after that. _____

10. Members of the school board survey 25 elementary school students, 25 middle school students, and 25 high school students. _____

Name _____ Date _____ Class _____

LESSON 7-8 Problem Solving
Populations and Samples

Write the correct answer.

1. Max wants to find out the exercise habits of local children. He plans to survey every third child he sees coming out of a sporting goods store. Max says his sample is not biased. Do you agree? Explain your answer.

_____

_____

2. Ms. Constantine is choosing among three field trips for her two classes. She wants to determine which trip her students prefer. Should she survey the entire population or use a sample? Explain.

_____

_____

3. A researcher catches 60 fish from different locations in a lake. He then tags the fish and puts them back in the lake. Two weeks later, the researcher catches 40 fish from the same locations. 8 of these 40 fish are tagged. Predict the number of fish in the lake.

_____

4. A high school has 1,800 students. A random sample of 80 shows that 24 have cell phones. Predict the number of students in the high school who have cell phones.

_____

Choose the letter for the best answer.

5. The school board wants to study computer literacy among teachers. Which would represent a random sample of teachers?

   A all high school math teachers
   B teachers from the middle school whose name begins with N
   C all male teachers
   D every eighth teacher on an alphabetical list

6. In a random sample, 3 of 400 computer chips are found to be defective. Based on the sample, about how many chips out of 100,000 would you expect to be defective?

   F 750
   G 3,000
   H 4,000
   J cannot be determined

Name _____ Date _____ Class _____

LESSON 7-8 Reading Strategies
Compare and Contrast

To get information about issues, a survey is conducted. Surveys can be done in two different ways.

- **Population** The entire group is surveyed.
- **Sample** Part of the entire group is surveyed.

1. Compare the difference between collecting information from the population and collecting a sample.

_____

There are two different types of samples.

- **Unbiased sample** The sample represents the population.
- **Biased sample** The sample does not represent the population.

2. What is the difference between an unbiased sample and a biased sample?

_____

_____

Mrs. Jones wants to know which sport 7th graders in the district like best. There are 7th graders in 6 different schools in the district. She can collect data in one of the following ways:

Population—Ask every 7th grade student at all 6 schools.
Unbiased sample—Ask every other 7th grader at 3 of the schools.
Biased sample—Ask 7th grade boys at 3 of the schools.

Write "unbiased sample" or "biased sample" **to describe each survey.**

3. A survey conducted at an ice cream store asked only mothers their favorite ice cream flavor.

_____

4. A reporter asked every tenth person coming out of a theater how well they liked the movie.

_____

5. A survey asked only girls to identify their favorite item on the school cafeteria menu.

_____

Name _____ Date _____ Class _____

Puzzles, Twisters, & Teasers
LESSON 7-8 *Answer This!*

Choose the best description of each sample.
Circle the letter above your answer.

1. Tyler surveys 60 people at a high school football game to find out which high school sports team is most popular.

   O biased   **A** not biased

2. A computer manufacturer randomly calls 150 people who called its help line and surveys them to determine if they were satisfied with the help they received.

   M biased   **N** not biased

3. Margie surveys 12 of her friends to determine what percentage of students participate in after-school clubs.

   F random sample   **D** convenience sample

4. Tamira surveys 50 riders on one commuter train to determine whether riders on all trains in the system think that the on-time performance is adequate.

   K biased   **A** not biased

5. Sam calls every sixth student on a high school attendance list and surveys the students to determine their favorite subject.

   M random sample   **R** convenience sample

Write the circled letters above the problem numbers to solve the riddle.

What is the difference between dinosaurs and dragons?

Dinosaurs ___ ___ ___ ___ T S ___ ___ ___ E
        3. 1. 2. 1.   5. 1. 4.

Name _____ Date _____ Class _____

LESSON 7-9 Practice A
Scatter Plots

The table shows how much energy was produced from wind power in the United States from 2000 to 2004.

Year	2000	2001	2002	2003	2004
Wind Power (trillions of BTUs)	57	70	105	115	143

1. Make a scatterplot of the data.

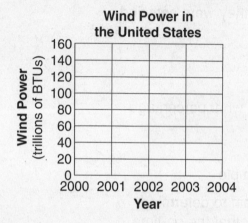

2. What does the scatter plot show about how much wind power was produced in the United States from 2001 to 2004?

_____

_____

3. What kind of correlation does the scatter plot show—positive, negative, or no correlation?

_____

Write *positive, negative,* or *no correlation* to describe each relationship.

4.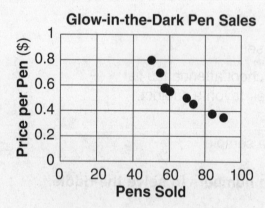

5.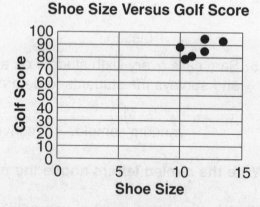

_____

6. outside temperature and the chance of your getting frostbite

_____

7. number of hours of spent studying for a test and the test scores

_____

8. number of basketballs going into the basket and distance in feet from the basket when shooting

_____

9. height and age of adults over 25

_____

Name _____ Date _____ Class _____

LESSON 7-9 Practice B
Scatter Plots

The table shows boys' average heights in inches from ages 6 through 13. Use the table for Exercises 1–3.

Age	6	7	8	9	10	11	12	13
Height (in.)	$46\frac{3}{4}$	49	51	$53\frac{1}{4}$	$55\frac{1}{4}$	$57\frac{1}{4}$	59	61

1. Make a scatter plot of the data.

2. Describe the relationship between the data sets.

   _____

3. What kind of correlation does the plot show?

   _____

Write *positive*, *negative*, or *no correlation* to describe each relationship.

4.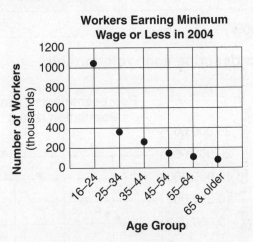

5.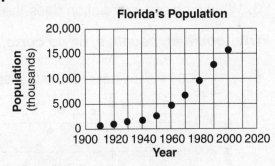

_____ _____

6. student test scores and the number of students who walk to school

7. the grade levels of students and their ages in months

_____ _____

8. the year a state entered the union and the number of years as a state

9. ages of students and their grades on tests

_____ _____

Name _____ Date _____ Class _____

LESSON 7-9 Practice C
Scatter Plots

The table shows the percentage of homes with a PC based on household income in 2001. Use the table for Exercises 1–3.

1. Make a scatter plot of the data.

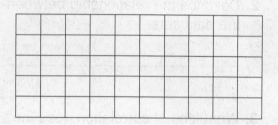

Annual Income	Homes with PCs
Less than $5,000	26%
$5,000–$9,999	19%
$10,000–$14,999	26%
$15,000–$19,999	32%
$20,000–$24,999	40%
$25,000–$34,999	50%
$35,000–$49,999	64%
$50,000–$74,999	78%
$75,000 and up	89%

2. Describe the relationship between the data sets.

_____

_____

3. What kind of a correlation does the plot show? _____

Write *positive*, *negative*, or *no correlation* to describe each relationship.

4.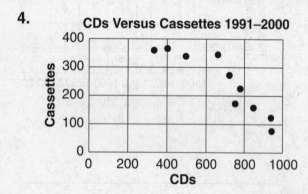

5.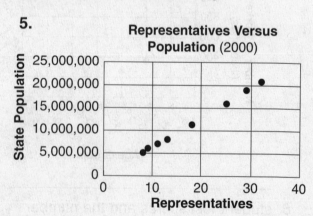

6. the number of senators and the state population

7. the number of hours spent exercising and the number of calories burned

Name _____ Date _____ Class _____

Reteach
CHAPTER 7-9 Scatter Plots

A **scatter plot** can be used to explore how two sets of data are related.

A scatter plot can show three different relationships between data sets.

Positive Correlation
y-axis values increase as x-axis values increase.

Negative Correlation
y-axis values decrease as x-axis values increase.

No Correlation

Look at each scatter plot. What kind of correlation is shown?

1. Feeding a Family

Weekly Grocery Bill vs. Number of People in Family

_____

2. Science and Your Feet

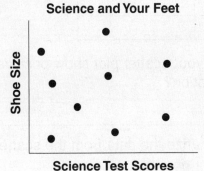

Shoe Size vs. Science Test Scores

_____

What type of correlation might you expect between:

3. the number of pages in a book and the number of copies the book sells?

_____

4. the number of computers per school and the number of students per computer?

_____

5. the number of students in a city and the number of students who play team sports?

_____

6. the temperature outside and the amount of heat used inside?

_____

7. a person's height and his or her birthday?

_____

8. the speed of a car and the time required to travel a certain distance?

_____

Copyright © by Holt, Rinehart and Winston.
All rights reserved.

73

Holt Mathematics

Name _____ Date _____ Class _____

LESSON 7-9

Challenge
Time On Task

Use a stopwatch to find out how long it takes to write your first and last names backwards. Do it 10 times, filling in the table with the time of each try. Use the data to make a scatter plot.

Try	Time
1	
2	
3	
4	
5	
6	
7	
8	
9	
10	

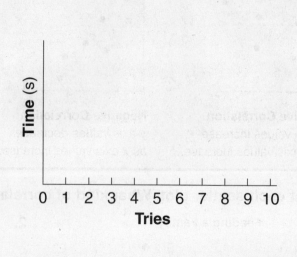

1. Does your scatter plot show positive, negative, or no correlation?

   _____

2. Summarize the data from the scatter plot.

   _____

   _____

3. Check the tasks that would probably have a scatter plot that is similar to the one you drew above after doing each task 10 times.

Making tacos in a fast food restaurant		Writing multiplication facts from 1 to 12	
Filling boxes on an assembly line		Eating dinner	
Driving from home to school		Talking to a friend on the phone	

4. Based on the data and scatter plot, what conclusion can you make about learning to do a job?

   _____

   _____

Name _____ Date _____ Class _____

LESSON 7-9 Problem Solving
Scatter Plots

Write the correct answer.

This scatter plot compares the mean annual income of Americans with the number of years spent in school.

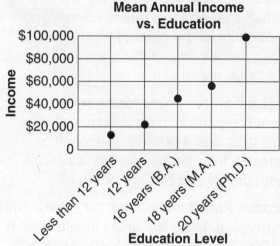

1. Which level of education has a mean annual income between $40,000 and $50,000?

   _____

2. Estimate the range of income data on this scatter plot.

   _____

3. Which level of education has the lowest income?

   _____

4. Does the scatter plot show a positive correlation, negative correlation, or no correlation between education and income?

   _____

Choose the letter for the best answer.

5. What kind of correlation would you expect to find between a city's annual snowfall amount and the size of its population?

   A positive correlation
   B negative correlation
   C no correlation
   D impossible to say

6. What kind of correlation would you expect to find between a movie's length and the number of times it can be shown in a day?

   F positive correlation
   G negative correlation
   H no correlation
   J impossible to say

7. What kind of correlation would you expect to find between an animal's mass and the number of calories it consumes in a day?

   A positive correlation
   B negative correlation
   C no correlation
   D impossible to say

8. What kind of correlation would you expect to find between a person's height and his or her income?

   F positive correlation
   G negative correlations
   H no correlation
   J impossible to say

Holt Mathematics

Name _____ Date _____ Class _____

LESSON 7-9

Reading Strategies
Drawing Conclusions

You can place individual data points on a graph to see if a pattern occurs. This type of graph is called a **scatter plot**.

Scatter Plot A identifies a car's age and its mileage. Age in years is listed along the bottom of the graph. Mileage is listed along the left side of the graph in intervals of 10,000 miles.

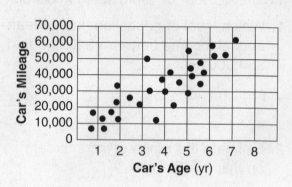

Notice how the points for each data pair create a graph with an upward pattern. This indicates a positive relationship between the number of miles a car has been driven and the age of the car.

Scatter Plot B identifies a car's value and its mileage. Miles are given in intervals of 10,000. Values are in intervals of $5,000.

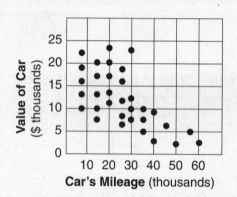

Notice how the points for each data pair create a downward pattern. This shows that the greater the mileage, the less a car is worth.

Scatter Plot C identifies a car's age and the number of passengers in the car.

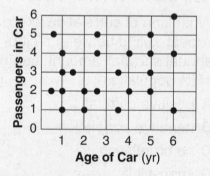

Notice how the points are scattered on the graph. This shows that there is no relationship between these two sets of data.

Read each statement. Identify the scatter plot that helps you determine the answer, and write "true" or "false."

1. The older the car, the fewer the miles. _____

2. The more miles on a car, the less its value. _____

3. The newer the car, the fewer the miles.
   _____

4. The older the car, the more passengers it can carry. _____

Copyright © by Holt, Rinehart and Winston.
All rights reserved.

76

Holt Mathematics

Name _____ Date _____ Class _____

LESSON 7-9 Puzzles, Twisters & Teasers
Scatter Brained!

Does the size of the brain determine the size of the body? Chart the data from the table below as a scatter plot. Then answer the questions.

Animal	Brain Weight (oz)	Body Weight (lb)
Human	53	140
Bottlenose Dolphin	56	340
Asian Elephant	263	10,000
Killer Whale	197	12,000
Cow	17	1,000
Mouse	0.014	0.024

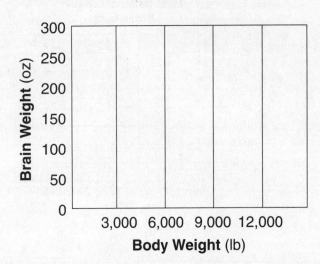

1. What kind of correlation do you see in the scatter plot?
   A positive c̲orrelation
   B negative correlation
   C no correlation

2. A mouse's brain makes up 3.2% of its body weight. A human's brain makes up 2.1% of its body weight. A cow's brain makes up 0.1% of its body weight. Which animal has the biggest brain in relation to its size?
   A mouse
   B human
   C co̲w

3. Which animal is the smartest, a cow, a human, or a bottlenose dolphin?
   A c̲ow
   B hu̲man
   C bottle̲nose dolphin

4. Based on your conclusions, does a bigger brain make a bigger body?
   A ca̲n't tell
   B no̲
   C sure̲

Use the underlined letters to solve this brain teaser.

How do you count a herd of cows?

With a ___ ___ ___ ___ ___ ___ ___ ___ ___ ___.

77   Holt Mathematics

Name _____ Date _____ Class _____

LESSON 7-10 Practice A
Misleading Graphs

1. Which graph could be misleading? Why?

Graph A

Graph B

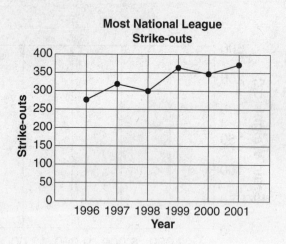
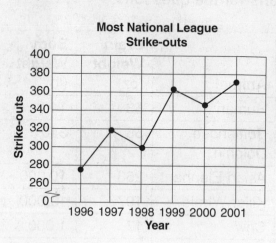

_____

Tell why each graph could be misleading.

2.

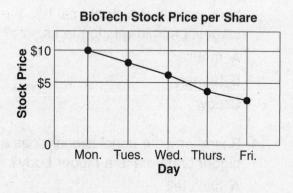

3.

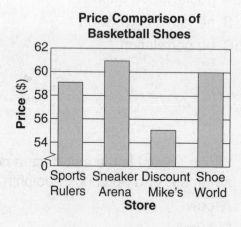

_____

_____

Name _____ Date _____ Class _____

LESSON 7-10
Practice B
Misleading Graphs

1. Which graph could be misleading? Why?

Graph A

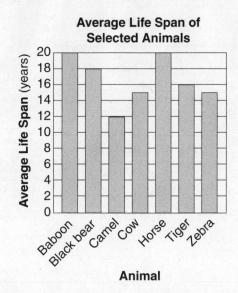

Graph B

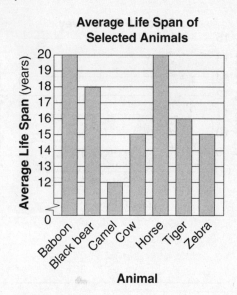

_____
_____

Explain why each graph could be misleading.

2.

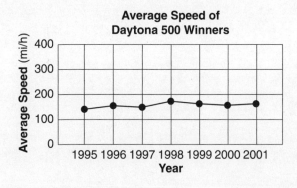

3.

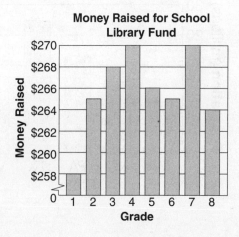

_____
_____
_____

Copyright © by Holt, Rinehart and Winston.
All rights reserved.

79

Holt Mathematics

Name _____ Date _____ Class _____

LESSON 7-10
Practice C
Misleading Graphs

1. Which graph could be misleading? Why?

Graph A

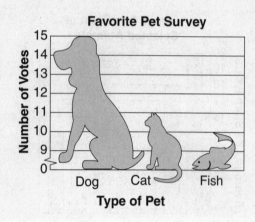

Graph B

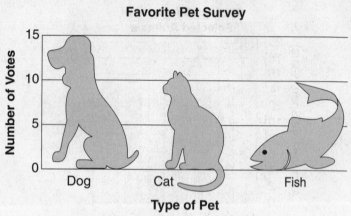

_____

_____

Explain why this graph could be misleading. Then redraw it so it is not misleading.

2.

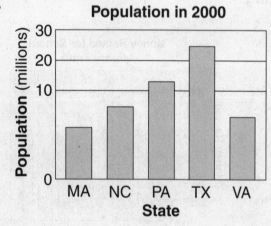

_____

Name _____ Date _____ Class _____

Reteach
LESSON 7-10 Misleading Graphs

Here are some ways a graph may be misleading.
- An axis is "broken" or numbers are left out on an axis.
- The intervals on an axis are not the same length.
- Different sizes are used to represent bars in a bar graph.
- Pictorial graphs could distort the data.

Compare these graphs.

It looks like sales dropped by more than half from January to February. Look at the vertical axis. It is broken, so the scale does not start at 0. This graph is misleading.

Look at the scale on the vertical axis. All the numbers from 0 to 1,000 are represented. This is a more realistic graph.

Explain why each graph is misleading.

1. Roller Skate Sales

2. Oceans of the World
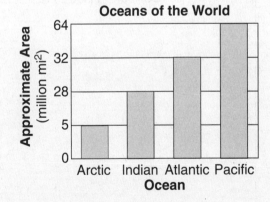

_____
_____
_____

_____
_____
_____

Copyright © by Holt, Rinehart and Winston.
All rights reserved.

Holt Mathematics

Name _____ Date _____ Class _____

Challenge
LESSON 7-10 *Picture the Headlines*

Read each fact. Draw a bar or line graph to support the two different headlines that report the fact.

1. **Fact:** In 2002, property taxes were $2.50 per $100 of the value of a house in Oakville. In 2003, they will increase to $2.60 for each $100 of value.

Property Taxes Skyrocket Under New Mayor!	**NEW MAYOR OKAYS SMALL INCREASE IN PROPERTY TAXES**

2. **Fact:** From 1996 to 1997, public schools in the United States averaged 7.8 students per computer. From 1997 to 1998, this average went down to 6.1. From 1998 to 1999, it was 5.7; and from 1999 to 2000, it was 5.4

Schools See Big Drop in Number of Students per Computer Since 1996	*Little Change Since 1996 in Number of Students per Computer in Our Public Schools!*

Copyright © by Holt, Rinehart and Winston.
All rights reserved.

Holt Mathematics

Name __Name_____ Date __Date__ Class __Class__

Problem Solving
LESSON 7-10 Misleading Graphs

Write the correct answer. Use the line graph for Exercises 1–4.

1. What would be a less misleading title for this graph?

   _____

   _____

2. How is the horizontal axis misleading?

   _____

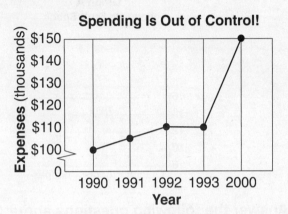

3. How is the vertical axis misleading?

   _____

4. How much did spending really increase between 1993 and 2000?

   _____

Choose the letter for the best answer.

The bar graph is an advertisement used by a tour company to convince New Yorkers to vacation in Hawaii.

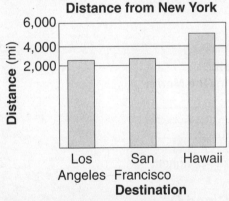

5. How far is New York from Los Angeles?
   A 2,500 miles   C 5,000 miles
   B 4,000 miles   D 5,200 miles

6. How far is New York from Hawaii?
   F 2,500 miles   H 5,000 miles
   G 2,600 miles   J 6,000 miles

7. What is the point of the ad?
   A New York is closer to Hawaii than to Los Angeles.
   B Hawaii is the same distance from New York as Los Angeles.
   C Hawaii is only slightly farther from New York than is Los Angeles.
   D San Francisco and Los Angeles are the same distance from New York.

8. Why is the graph misleading?
   F The distances are incorrect.
   G The bars are mislabeled.
   H The bars are too tall.
   J The intervals on the vertical axis are not equal.

Holt Mathematics

Name _____ Date _____ Class _____

Reading Strategies
LESSON 7-10 Analyze Information

Graphs can be misleading if the data is not shown properly.

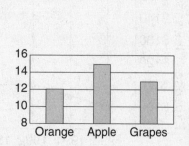

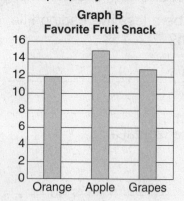

Answer the following questions about Graphs A and B.

1. What number does the scale start on for Graph A? _____

2. What number does the scale start on for Graph B? _____

3. Which of the graphs is misleading? Why?
   _____
   _____

A graph is misleading if the intervals are too large or too small.

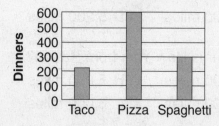

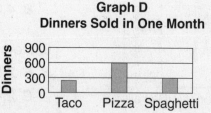

Answer the following questions about Graphs C and D.

4. What intervals are used in Graph C? _____

5. What intervals are used in Graph D? _____

6. Which graph is misleading? Why?
   _____
   _____

Name _____ Date _____ Class _____

LESSON 7-10 Puzzles, Twisters & Teasers
How's Your Vocabulary?

Complete the crossword puzzle by answering the clues.

Across

2. Each _____ or slice of a circle represents one part of the entire data set.

5. Each point on a scatter _____ represents a pair of data values.

7. bar or line _____

8. Frequency and cumulative frequency are two types of _____.

11. The _____ is the value or values that occur most often in a data set.

12. A _____ graph shows change over time for two sets of data.

13. The _____ is the sum of the data values divided by the number of data items.

14. The difference between the least and greatest values is called the _____.

Down

1. Often researchers can't survey every member of a large group, so they study a part of the group. This group is called a _____.

3. Positive, negative, or no _____ are three ways to describe data in a scatter plot.

4. the singular form of data

6. An extreme value in a set is called an _____.

9. The _____ is the middle value of an odd number of items arranged in order and it is the average of the two middle values for an even number of items.

10. A _____ table is one way to organize data into categories or groups.

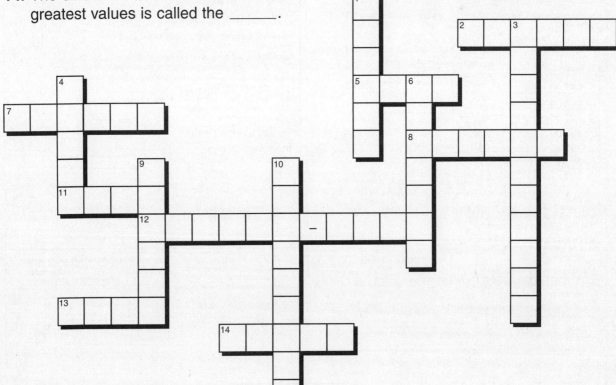

85
Holt Mathematics

Practice A
7-1 Frequency Tables, Stem-and-Leaf Plots, and Line Plots

The table shows normal monthly temperatures for Tampa, Florida, for each month of the year.

Normal Monthly Temperatures in Tampa

Month	January	February	March	April	May	June
Temp. (°F)	61	63	67	72	78	82

Month	July	August	September	October	November	December
Temp. (°F)	83	83	82	76	69	63

1. Complete the cumulative frequency table for the data.

Normal Monthly Temperatures in Tampa

Temperature (°F)	Frequency	Cumulative Frequency
60-69	5	5
70-79	3	8
80-89	4	12

2. How many months had a normal temperature of less than 80°? __8 months__

3. Complete a stem-and-leaf plot of the data.

Normal Monthly Temperatures in Tampa

Stems	Leaves
6	1 3 3 7 9
7	2 6 8
8	2 2 3 3

Key: 6 | 1 means 61

4. How many months had a normal temperature in the 60s? __5 months__

5. Complete a line plot of the data.

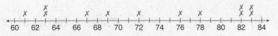

Practice B
7-1 Frequency Tables, Stem-and-Leaf Plots, and Line Plots

The table shows the heights of students in Ms. Blaire's class. Use the table for Exercises 1 and 2.

	Height (in.)
Males	60, 45, 48, 57, 62, 59, 57, 60, 56, 58, 61, 52, 55
Females	49, 52, 56, 48, 51, 60, 47, 53, 55, 58, 54

1. Make a cumulative frequency table of the data.

Heights of Students

Height (in.)	Frequency	Cumulative Frequency
45–49	5	5
50–54	5	10
55–59	9	19
60–64	5	24

2. How many of the students were less than 60 in tall? __19__

3. Make a stem-and-leaf plot of the data.

Height of Students

Stem	Leaves
4	5 7 8 8 9
5	1 2 2 3 4 5 5 6 6 7 7 8 8 9
6	0 0 0 1 2

Key: 5 | 2 means 52

4. How many of the students were less than 50 in tall? __5__

5. Make a line plot of the data.

6. Which height occurred the greatest number of times? __60 in.__

Practice C
7-1 Frequency Tables, Stem-and-Leaf Plots, and Line Plots

The list below shows the ages of all the presidents inaugurated in the twentieth century.
42, 51, 56, 55, 51, 54, 51, 60, 62, 43, 55, 56, 61, 52, 69, 64, 46

1. Make a cumulative frequency table of the data.
2. Make a stem-and-leaf plot of the data.

Ages of Twentieth Century Presidents

Ages	Frequency	Cumulative Frequency
40-44	2	2
45-49	1	3
50-54	5	8
55-59	4	12
60-64	4	16
65-69	1	17

Ages of Twentieth Century Presidents

Stems	Leaves
4	2 3 6
5	1 1 1 2 4 5 5 6 6
6	0 1 2 4 9

Key: 4 | 6 means 46

3. Make a line plot of the data.

4. How many of these presidents were in their 50s when they took office? __9__

5. Which of the following is the most likely source of the data in the stem-and-leaf plot? Explain your answer.

	Stems	Leaves
	5	1 1 2 2 3 4 5 6 8 8
	6	0 0 0 0 4 1 8
	7	1 2

Key: 5 | 8 means 58

a. class sizes in a middle school
b. speeds in miles per hour of cars on a highway
c. number of hours worked in a week by part-time employees of Toy Town

Possible answer: b; Cars on a highway could be traveling between 51 and 72 miles per hour. Classes probably would not have 51-72 students, and part-time employees would not be working that many hours.

Reteach
7-1 Frequency Tables, Stem-and-Leaf Plots, and Line Plots

A **frequency table** organizes data into groups.

A **cumulative frequency** table includes a running total of the frequencies of all the previous groups.

The list shows the pulse rate of 12 students.
65, 63, 72, 88, 90, 82, 80, 75, 68, 72, 74, 84

You can use the data to make a cumulative frequency table. Here's how.

Step 1: Choose an interval. Each interval must be the same size. Often, 5 or 10 is a good interval length. Write the intervals in the first column.

Step 2: Find the number of values in each interval. Write the number in the middle column.

Step 3: Add the values to find the cumulative frequency. Write the number in the last column.

Pulse Rates

Pulse Rate	Frequency	Cumulative Frequency
60–69	3	3
70–79	4	4 + 3 = 7
80–89	4	4 + 7 = 11
90–99	1	1 + 11 = 12

intervals of 10

the number of students with pulse rates between 80 and 89

the sum of the frequency column is equal to the last number in the cumulative frequency column.

1. The list below shows the highest temperatures (°F) ever recorded in all the states that begin with the letters M or N in the United States. Use the data to complete the cumulative frequency table.
105°, 109°, 107°, 112°, 114°, 115°, 118°, 117°, 118°, 125°, 106°, 110°, 122°, 108°, 110°, 121°

Highest Temperatures in M and N States

Temperature (°F)	Frequency	Cumulative Frequency
105–109	5	5
110–114	4	9
115–119	4	13
120–124	2	15
125–129	1	16

LESSON 7-1 Reteach
Frequency Tables, Stem-and-Leaf Plots, and Line Plots (continued)

A **stem-and-leaf plot** shows how often data values occur and how they are distributed. Numbers are separated into two parts: the stem and the leaves. The stem shows the leftmost digit in each number, and the leaves show the rightmost digit in each number.

You can use the pulse rate data to make a stem-and-leaf plot. Here's how.

65, 63, 72, 88, 90, 82, 80, 75, 68, 72, 74, 84

Step 1: List the stems in the left column of the plot.
Step 2: List the leaves in order in each row that corresponds to the stem of the number.
Step 3: Write a key and a title.

Pulse Rates

Stems	Leaves
6	3 5 8
7	2 2 4 5
8	0 2 4 8
9	0

Key: 7 | 2 means 72

The stems are the left digits of each number. The leaves are the right digits of each number.

The first stem (6) has three leaves (3, 5, and 8). This represents the data values 63, 65, and 68.

2. The list below shows the number of stories of some of the tallest buildings in Houston, Texas. Use the data to complete the stem-and-leaf plot.
55, 52, 53, 64, 56, 71, 50, 46, 49, 75
50, 49, 46, 36, 47, 45, 47, 36, 44, 42

Tallest Buildings in Houston

Stems	Leaves
3	6 6
4	2 4 5 6 6 7 7 9 9
5	0 0 2 3 5 6
6	4
7	1 5

Key: 3 | 6 means 36

LESSON 7-1 Challenge
Read From the Middle Out

/27

In a double stem-and-leaf plot, the stem is in the middle and the leaves are on both sides. You read from the middle to the left for the left data and the middle to the right for the right data.

The double stem-and-leaf plot compares the average monthly temperatures in Bloomington, Indiana, and Richmond, Virginia, in degrees Fahrenheit.

Average Monthly Temperatures (°F)

Bloomington Leaves	Stems	Richmond Leaves
7	2	
3 1	3	8
5 2	4	0 0 8 9
6 3	5	7 9
7 3	6	6
6 4 2	7	0 4 7 8

Key: 7 | 2 means 27°F
Key: 3 | 8 means 38°F

Use the double stem-and-leaf plot above to answer the questions. **Possible answers are given.**

1. What is the greatest average monthly temperature in Bloomington? in Richmond?
76°F; 78°F

2. Which city had the lowest monthly temperature?
Bloomington

3. Which city has more months with monthly temperatures below 30°F?
Bloomington

4. Which city has more variation in temperatures? Explain.
Bloomington; It has more low temperatures. Greater Range

5. What average temperature did Richmond have twice?
40°F

6. How many months was the average monthly temperature above 68° in Richmond?
4 months

7. What was the lowest average monthly temperature in Bloomington?
27°F

8. Compare the highest average monthly temperature in Bloomington to the highest average monthly temperature in Richmond.
Richmond's highest average temperature was 2° higher than Bloomington's.

/10

LESSON 7-1 Problem Solving
Frequency Tables, Stem-and-Leaf Plots, and Line Plots

Write the correct answer.
The table shows the time in minutes that Naima talked on the phone during the last 3 weeks.

Phone Time (min)

	Mon	Tues	Wed	Thurs	Fri	Sat	Sun
Week 1	12	15	25	45	52	30	31
Week 2	22	25	46	51	10	19	33
Week 3	44	21	30	20	10	24	52

1. Naima made a cumulative frequency table of the data using equal intervals. What number would she write in the frequency column for the interval 11–20 minutes?
4

2. Naima made a line plot of the data. Which numbers had more than one X above them?
10, 25, 30, 52

3. If Naima makes a stem-and-leaf plot, which stem has the most leaves? What are they?
stem 2; leaves 0, 1, 2, 4, 5, 5

4. In the stem-and-leaf plot, which stems have the same number of leaves?
Stems 4 and 5 each have 3 leaves.

The list shows Hank Aaron's season home run totals. Make a cumulative frequency table, stem-and-leaf plot, and a line plot for the data. Then use the data to solve problems 5 – 8.

13, 27, 26, 44, 30, 39, 40, 34, 45, 44, 24, 32
44, 39, 29, 44, 38, 47, 34, 40, 20, 12, 10

5. In a cumulative frequency table of the data, what number belongs in the frequency column for interval 40-44?
A 5 C 8
B 6 D 14

6. In a cumulative frequency table of the data, what number belongs in the frequency column for interval 25-29?
F 3 H 6
G 5 J 8

7. In a stem-and-leaf plot of the data, how many stems do you need?
A 1 C 3
B 2 **D 4**

8. In a line plot of the data, which number would have 4 x's above it?
F 34 **H 44**
G 40 J 45

LESSON 7-1 Reading Strategies
Use Graphic Aids

To the right is a listing of points scored in each of the league's basketball games.

28 48 34 50 47 35 40 37 36
55 43 39 43 34 52

Creating a **frequency table** is one way to organize the scores. A frequency table organizes scores by how often they occur.

League Basketball Scores

Points Scored	Frequency
20–29	1
30–39	6
40–49	5
50–59	3

Answer these questions about the frequency table.

1. How are the scores organized?
in intervals of 10

The **frequency**, or how often the scores occur, is recorded in the second column.

2. In how many games were 50 or more points scored?
3 games

3. In how many games were between 30 and 39 points scored? **6** games

A **stem-and-leaf plot** is another way to organize the data. Since the scores are two-digit numbers, the **stem** represents the tens digits, and the **leaves** represent the ones in each score.

Data in this row: 50, 52, 55

League Basketball Scores

Stems (tens)	Leaves (ones)
2	8
3	4 4 5 6 7 9
4	0 3 3 7 8
5	0 2 5

Key: 2 | 8 means 28

Answer each question about the stem-and-leaf plot.

4. Are the stems on the right or the left in a stem-and-leaf plot? **left**
5. How many stems are in this plot? **4 stems**
6. What number does 3|6 represent? **36**

/4

LESSON 7-1 Puzzles, Twisters & Teasers
The Plot Thickens!

Have you ever wondered about traveling trees? Finish this page to discover the answers to two important traveling tree questions! The list shows the number of minutes people spend watching television each day. Fill in the blanks to make a cumulative frequency table of the data.

10, 10, 15, 18, 25, 28, 32, 35, 35, 45, 50, 55, 57, 59

Minutes	Frequency	Cumulative Frequency
10–20	4	4 A
21–30	2	6 E
31–40	3	9 L
41–50	2	11 S
51–60	3	14 V

What does a tree do when it wants to go away?

It L   E   A   V   E   S
   9   6   4 14   6 11

Now use the list to make a stem-and-leaf plot.

Stems	Leaves	
1	0 0 5 8	K
2	5 8	N
3	2 5 5	R
4	5	T
5	0 5 7 9	U

Where does a traveling tree pack its things?

In its T   R   U   N   K
       45 32 57 28 15

LESSON 7-2 Practice A
Mean, Median, Mode, and Range

Find the mean, median, mode, and range.

1. 5, 8, 5, 9, 3
   6; 5; 5; 6

2. 58, 54, 60, 56, 52
   56; 56; no mode; 8

3. 18, 17, 21, 18, 26
   20; 18; 18; 9

4. 60, 20, 40, 10, 50, 30
   35; 35; no mode; 50

The line plot below shows the number of kilometers Clara ran each day for 14 days. Use the line plot for Exercises 5 and 6.

5. Find the mean, median and mode for the set of data.
   7; 5.5; 5 and 6

6. Which measure of central tendency best describes the data? Explain your answer.
   Possible answer: The median best describes the data set because it is closer to the number of kilometers that Clara ran the majority of the days than the mean is. There are two modes

Use the data set to answer the questions.
22, 18, 15, 19, 61, 21   15, 18, 19, 21, 22

7. What is the outlier? 61

8. How does the outlier affect the mean, median, and mode?
   It increases the mean by 7 and the median by 1. It does not affect the mode, because there is no mode.

9. Which measure of central tendency best describes the data set with the outlier?
   Possible answer: The median best describes the data set with the outlier.

10. Which measure of central tendency best describes the data set without the outlier?
    Possible answer: The mean best describes the data set without the outlier.

LESSON 7-2 Practice B
Mean, Median, Mode, and Range

Find the mean, median, mode, and range of each data set.

1. 46, 35, 23, 37, 29, 53, 43
   38; 37; no mode; 30

2. 72, 56, 47, 69, 75, 48, 56, 57
   60; 56.5; 56; 28

3. 19, 11, 80, 19, 27, 19, 10, 25, 15
   25; 19; 19; 70

4. 7, 8, 20, 6, 9, 11, 10, 8, 9, 8
   9.6; 8.5; 8, 14

5. The line plot shows the number of hours 15 students said they spent on homework in one week. Which measure of central tendency best describes the data? Justify your answer.

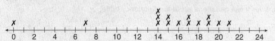

The median best describes data set because it is closest to the numbers of hours reported by most of the students.

Identify the outlier in each data set. Then determine how the outlier affects the mean, median, and mode of the data. Then tell which measure of central tendency best describes the data with and without the outlier.

6. 14, 16, 13, 15, 5, 16, 12
   Outlier, 5; It decreased the mean by 1.3 and the median by 0.5. It did not affect the mode. With the outlier, the data is best described by the median. Without the outlier, the data is best described by the mean.

7. 48, 46, 52, 92, 57, 58, 52, 61, 56
   Outlier, 92; It increased the mean by 4.25 and the median by 2. It did not affect the mode. With the outlier, the data is best described by the median. Without the outlier, the data is best described by the mean.

LESSON 7-2 Practice C
Mean, Median, Mode, and Range

Find the mean, median, mode, and range of each data set.

1. 21, 14, 17, 16, 23, 17, 15, 13
   17; 16.5; 17; 10

2. 44, 56, 38, 55, 62, 56, 48, 62, 56
   53; 56; 56; 24

3. The line plot shows Omar's point totals for 20 basketball games. Which measure of central tendency best describes the data? Justify your answer.

Possible answer: The median best describes data set because it is closest to the majority of the point totals.

Identify the outlier in each data set. Then determine how the outlier affects the mean, median, and mode of the data. Then tell which measure of central tendency best describes the data with and without the outlier.

4. 26, 33, 36, 26, 30, 26, 10, 33
   Outlier, 10; It decreased the mean by 2.5 and the median 2. It did not affect the mode. With the outlier, the data is best described by the median. Without the outlier, the data is best described by the mean.

5. 59, 56, 62, 103, 61, 67, 62, 52, 63
   Outlier, 103; It increased the mean by 4.75 and the median by 0.5. It did not affect the mode. With the outlier, the data is best described by the median. Without the outlier, the data is best described by the mean.

6. Bart has these quiz scores: 78, 76, 74, 80, 82, 78, 82, 75, 77, 82. Bart tells his parents that his typical quiz score is an 82. Which measure of central tendency did Bart use to describe his typical score? Do you think Bart's measure of central tendency best described his typical score? Explain.
   mode; Possible answer: No; the mode was higher than all of the other scores, and occurred only one more time than the score that occurred the next greatest number of times.

LESSON 7-2 Reteach
Mean, Median, Mode, and Range

Measures of central tendency show what the middle of a data set looks like. The measures of central tendency are the *mean, median,* and *mode.*

Find the mean, median, mode, and range of 8, 3, 5, 4, 1, and 3.

- **Find the mean.** The mean is the sum of the values divided by the number of values in the data set.
  $1 + 3 + 3 + 4 + 5 + 8 = 24$
  $24 \div 6 = 4$
  mean = 4

- **Find the range.** Find the difference between the least and greatest values.
  $8 - 1 = 7$
  range = 7

- List in order: 1, 3, 3, 4, 5, 8

- **Find the mode.** The mode is the value that occurs most often. Sometimes there is no mode.
  mode = 3

- **Find the median.** The median is the middle value.
  median = 3.5

Find the range, mean, median, and mode of each data set.

1. 6, 5, 3, 6, 8   5; 5.6; 6; 6
2. 12, 15, 17, 9, 17   8; 14; 15; 17
3. 26, 35, 23, 27, 19, 23   16; 25.5; 24.5; 23
4. 7, 6, 13, 16, 15, 9   10; 11; 11; no mode
5. 42, 38, 45, 42, 43   7; 42; 42; 42
6. 51, 62, 68, 55, 68, 62   17; 61; 62; 62 & 68

7. **Monthly Low Temperatures**

Month	Jun.	Jul.	Aug.	Sept.	Oct.
Temperature (°F)	44	41	47	42	36

11°F; 42°F; 42°F; no mode

LESSON 7-2 Reteach
Mean, Median, Mode, and Range (continued)

An **outlier** is a value that is much greater than or much less than the other values in a data set.

How does the outlier affect the mean, median, and mode of the data?
7, 9, 30, 9, 5, 6

Write the data in order and identify the outlier.
5, 6, 7, 9, 9, **30** ← outlier

With the Outlier	Without the Outlier
Find the mean. $7 + 9 + 30 + 9 + 5 + 6 = 66$ $66 \div 6 = 11$ The mean is 11.	Find the mean. $7 + 9 + 9 + 5 + 6 = 36$ $36 \div 5 = 7.2$ The mean is 7.2.
Find the median. 5, 6, 7, 9, 9, 30 $7 + 9 = 16$ $16 \div 2 = 8$ The median is 8.	Find the median. 5, 6, 7, 9, 9 The median is 7.
The mode is 9.	The mode is 9.

8. How does the outlier affect the mean?   It increases the mean by 3.8.
9. How does the outlier affect the median?   It increases the median by 1.
10. How does the outlier affect the mode?   There is no effect.

To choose the measure of central tendency that best describes a set of data:
- Choose the measure that is closest to the greatest number of values in the data set, OR
- If there is an outlier, think about how it affects the mean and the median. Choose the measure that is affected least by the outlier.

Use the data set to answer the questions.
4, 6, 3, 6, 25, 3, 2

11. Is there an outlier? If so, what is it?   Yes; 25
12. How does the outlier affect the mean and the median?
    It increases the mean by 3 and the median by 0.5.
13. Which measure of central tendency best describes the data? Explain your answer.
    The median best describes the data set because it is least affected by the outlier.

LESSON 7-2 Challenge
These Puzzles Are Mean

Solve each puzzle.

1. There are 6 whole numbers in a set of numbers. The least number is 8, and the greatest number is 14. The mean, the median, and the mode are 11. What are the numbers?
   Possible answer: 8, 10, 11, 11, 12, 14

2. There are 7 whole numbers in a set of numbers. The least number is 10, and the greatest number is 20. The median is 16, and the mode is 12. The mean is 15. What are the numbers?
   Possible answer: 10, 12, 12, 16, 17, 18, 20

3. There are 8 whole numbers in a set of numbers. The greatest number is 17, and the range is 9. The median and the mean are 12, but 12 is not in the data set. The modes are 9 and 14. What are the numbers?
   Possible answer: 8, 9, 9, 10, 14, 14, 15, 17

4. The mean of a data set of 6 numbers is 8. The mean of a different data set of 6 numbers is 20. What is the mean of the combined data sets?   14

5. Find the mean of 7 numbers if the mean of the first 4 numbers is 5 and the mean of the last 3 numbers is 12. What is the mean of the combined data sets?   8

6. The mean of a data set of 3 numbers is 12. The mean of a data set of 9 numbers is 40. What is the mean of the combined data sets?   33

7. Sasha needs an average of 30 points to move to the next level in her competition. Her scores in the first three events are 28, 35, and 30. What is the lowest score she can score in her last event to move to the next level of competition?   27

8. Lars has a score of 89 for each of his first 3 science quizzes. The score on his fourth quiz is 92. What score does he need on his fifth quiz to have an average of 90?   91

9. Jake has a 95 average in math after 4 quizzes. Then he got 0 on the next quiz after being absent. There are 2 more quizzes. What average grade does he need on these last 2 quizzes to keep his average at least 85?
   He can't do it. He would need to score about 108 on each quiz.

LESSON 7-2 Problem Solving
Mean, Median, Mode, and Range

Write the correct answer.

The table to the right shows the leading shot blockers in the WNBA during the 2003 season.

Player	Shots Blocked
Margo Dydek	100
Lauren Jackson	64
Lisa Leslie	63
Ruth Riley	58
Michelle Snow	62

1. What is the range of this set of data?
   42

2. What are the mean, median, and mode of this set of data?
   69.4; 63; there is no mode

3. What is the outlier in this set of data?
   100

4. How does the outlier affect the mean and the median?
   It increases the mean by 7.65 and the median by 0.5

5. Which measure of central tendency best describes the set of data with the outlier? Explain.
   Possible answer: The median; the outlier affects the median less than it does the mean.

Choose the letter for the best answer.

In a 100-meter dash, the first 5 racers finished with the following times: 11.6 seconds, 13.4 seconds, 10.8 seconds, 11.8 seconds, and 13.4 seconds.

6. Which measure of central tendency for this set of data is 12.2 seconds?
   A mean
   B median
   C mode
   D none of the above

7. Which measure of central tendency for this set of data is 11.8 seconds?
   F mean
   G median
   H mode
   J none of the above

8. What is the mode for this set of data?
   A 10.8 seconds
   B 11.8 seconds
   C 13.4 seconds
   D none of the above

9. The sixth racer finished with a time of 16.4 seconds. How will that affect the mean for this set of data?
   F decrease it by 0.7 second
   G increase it by 0.7 second
   H increase it by 3.28 seconds
   J does not affect the mean

Holt Mathematics

LESSON 7-2 Reading Strategies
Understanding Vocabulary

There are different ways to look at data. The **mean**, the **median**, the **mode**, and the **range** help us understand the information we gather.

The mean, or **average,** is the sum of the values divided by the total number of values. Example: Ages in years → 18 11 6 12 18

The average age is (18 + 11 + 6 + 12 + 18) ÷ 5 = 13 years

The **median** is the middle value in a set of data, when the values are listed in order. The median is 12.

6   11   **12**   18   18

The **mode** is the most frequent value in a set of data. The mode is 18.

6   11   **18**   **18**

The **range** indicates the spread of the data. It is the difference between the least and greatest values in a set of data.

18 − 6 = **12**. The range is 12.

These are Jack's test scores for 5 math tests: 83, 79, 92, 95, 83. Use this data to help Jack look at his scores. Write "mean," "median," "mode" or "range" to answer the following questions.

1. 83 occurs more than any of the other values. Which measure is 83?

   mode

2. Jack added his test scores and divided by 5 to find which measure?

   mean

3. Jack found that the difference between his highest and lowest score was 16 points. What is this measure called?

   range

4. Jack found that the middle value is 83. What is that value called?

   median

LESSON 7-2 Puzzles, Twisters & Teasers
Stuck in the Middle!

Ever wonder what number in a data set sticks out like a sore thumb? To find out, use the data sets below to answer each question. Fill in the letter corresponding to each question above the question's answer in the decoder below.

1, 4, 2, 9, 5, 3

T: What is the range? ___8___

L: What is the mean? ___4,___

I: What is the median? ___3.5___

6, 7, 4, 8, 12, 2, 1, 2

U: What is the range? ___11___

R: What is the mean? ___5.25___

O: What is the median? ___5___

E: What is the mode? ___2___

What number in some data sets sticks out like a sore thumb?

O	U	T	L	I	E	R
5	11	8	4	3.5	2	5.25

LESSON 7-3 Practice A
Bar Graphs and Histograms

The bar graph shows the lengths of four rivers. Use the graph for Exercises 1–3.

1. Which river is the longest?

   Nile River

2. About how much longer is the Amazon River than the Congo River?

   about 1,100 mi

3. About how much longer is the Nile River than the Huang River?

   about 700 mi

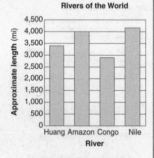

4. The table shows the ages of four U.S. Presidents when they first entered office. Make a bar graph of the data.

Name	President's Age
Truman	60
Kennedy	43
Carter	52
Bush	54

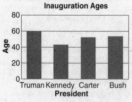

5. The table below shows the scores on a Spanish test. Make a histogram of the data.

Scores	Frequency
71–80	4
81–90	8
91–100	5

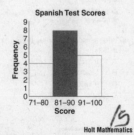

LESSON 7-3 Practice B
Bar Graphs and Histograms

The bar graph shows the elevations of the highest points in several states. Use the graph for Exercises 1–3.

1. Which state has the highest elevation?

   Alaska

2. About how much higher is Granite Peak than Guadalupe Peak?

   about 4,000 ft

3. About how much higher is Mount Whitney than Mount Marcy?

   about 9,000 ft

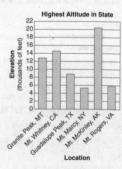

4. The table shows the approximate mean length and width of three states. Make a double-bar graph of the data.

State	Length (mi)	Width (mi)
Florida	500	160
New York	330	283
Virginia	430	200

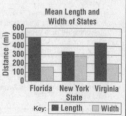

5. The list shows the bowling scores of the first game played by a group of bowlers on Thursday night. Make a histogram of the data. 96, 110, 132, 128, 105, 94, 116, 95, 126, 114, 123, 136, 121, 99

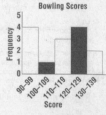

Practice C
7-3 Bar Graphs and Histograms

The bar graph shows the activities in which most injuries occur for children ages 5–14. Use the graph for Exercises 1–3.

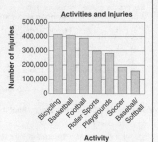

1. Which activity has the greatest number of injuries?
   bicycling

2. About how many more children are injured playing basketball than soccer?
   about 225,000 children

3. About how many more children are injured participating in roller sports than on playgrounds?
   about 20,000 children

4. The table shows the life expectancies of people living in three North American countries. Make a double-bar graph of the data.

Country	Male	Female
Canada	76	83
Greenland	66	74
Mexico	69	75

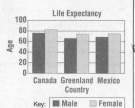

5. The list shows the times, in minutes, in which runners finished a 10-kilometer race. Make a histogram of the data.

   92, 97, 100, 88, 85, 79, 76, 83, 92, 87, 78, 85, 79, 98, 83, 84, 86

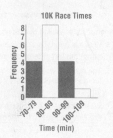

Reteach
7-3 Bar Graphs and Histograms

A **bar graph** uses bars to compare data.
A **double-bar graph** compares two sets of data.

Use the graph for Exercises 1–3. The bar graph shows the approximate amount of passenger traffic through some U.S. airports in 2004.

1. Which airport had more than 50 million arrivals and departures?
   Dallas

2. About how many arrivals and departures did Philadelphia have?
   about 29 million

3. About how many more people passed through the Denver airport than the Orlando airport?
   about 11 million

4. The table shows which sports are played by a seventh-grade class. Use the data in the table to make a double-bar graph.

Sport	Number of Boys	Number of Girls
Soccer	14	8
Tennis	7	10
Basketball	18	6
Swimming	12	20

a. Finish labeling the vertical axis.
   What is the vertical axis label?
   Number of Students

   What is the horizontal axis label?
   Sport

b. Draw a bar on the horizontal axis for the number of boys playing each sport. Draw a bar for the number of girls playing each sport. Label the bars.

   Each interval must be the same size.

c. Give the graph a title. Make a key.

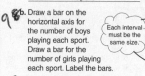

Reteach
7-3 Bar Graphs and Histograms (continued)

A **histogram** is used to represent data in a frequency table. In a histogram the bars touch and both axes show numerical data.

Use the histogram for Exercises 5–7.

The histogram shows the pulse rates of 20 patients during check-ups.

5. How many patients had a pulse rate between 80 and 89?
   5 patients

6. What range of pulse rates was most common among the patients?
   90–99

7. How many more patients had a pulse rate between 100 and 109 than between 60 and 69?
   3 more patients

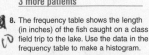

8. The frequency table shows the length (in inches) of the fish caught on a class field trip to the lake. Use the data in the frequency table to make a histogram.

Fish Caught on Class Trip

Length (in.)	Frequency
1–5	1
6–10	6
11–15	8
16–20	5
21–25	2

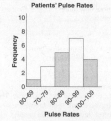

a. Finish labeling the horizontal axis.
   What is the vertical axis label? **Frequency**
   What is the horizontal axis label? **Fish Length (in.)**
b. The vertical axis shows the frequency. Finish labeling the axis.
c. Draw a bar for each interval. The bars should touch.
d. Give the graph a title.

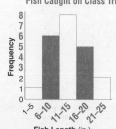

Challenge
7-3 Sliding Histogram

The graph below is called a sliding histogram. It shows Germany's population in 2005.

Use the sliding histogram to the right to answer the questions.

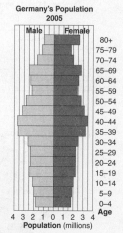

1. What is the length of the interval used on the histogram?
   5 years

2. Which axis shows the frequency?
   horizontal

3. Were there more males or females 85 years and older in Germany in 1997?
   females

4. About how many females 35–39 years old were there?
   about 3.3 million

5. About how many males 35–39 years old were there?
   about 3.6 million

6. Which age group had the greatest number of females? of males?
   40-44; 40-44

7. About what was the combined population of males and females 10–14 years old?
   about 4.3 million

8. About how many more females 75–79 were there than males 75–79?
   about 0.6 million

9. About how many more males were there 40–44 than 45–49?
   about 0.5 million

10. At what age does the number of females start to differ significantly from the number of males?
    age 75 and above

LESSON 7-3 Problem Solving
Bar Graphs and Histograms

Write the correct answer.

The double-bar graph shows the win-loss records for the Carolina Panthers football team in the years 1998-2004.

1. During how many seasons did the Panthers lose more games than they won?

   5 seasons

2. In which year did the Panthers win more games than they lost?

   2003

3. Between which 2 years did the Panthers have the greatest improvement in their win-loss record?

   between 2001 and 2002

4. In which year do you find the greatest range in the win-loss record?

   2001

Choose the letter for the best answer.

The histogram shows the ages of all members in a fan club.

5. How many fan club members could be teenagers?
   - **A** 5
   - **B** 8
   - **C** 17
   - **D** 21

6. How many fan club members are between the ages of 30 and 39?
   - **F** 5
   - **G** 8
   - **H** 17
   - **J** 22

7. In which situation would you use a histogram to display data?
   - **A** to show how you spend money
   - **B** to show the change in temperature throughout the day
   - **C** to show the golf scores from the whole team
   - **D** to show the life expectancy of different animals

8. In which situation would you use a bar graph to display data?
   - **F** to compare the speed of different computers
   - **G** to show how a cat spends its time
   - **H** to show how a child's height changes as he or she grows
   - **J** to show the distribution of math grades in your class

---

LESSON 7-3 Reading Strategies
Reading a Graph

Bar graphs create pictures for data.
A random sample was taken of boys' favorite sports.

Answer the following questions about the bar graph.

1. What does the graph show? __boys' favorite sports__
2. What number does the scale of the graph count by? __tens__
3. How many boys like basketball best? __about 25 boys__
4. How many boys like hockey best? __about 54 boys__

Coaches were surveyed about the number of hours their teams practice each month. The data is shown in a frequency table and in a **histogram**. A histogram displays data in equal intervals with connected bars.

Answer the following questions about the histogram.

5. Where is the number of coaches located on the histogram?
   along the left side of the graph

6. The numbers on the left side of the histogram are in equal groups of what number?
   10

7. What information is located along the bottom of the graph?
   the range of hours that teams practice each month

LESSON 7-3 Puzzles, Twisters & Teasers
A Graphic Display!

Answer the question by studying the bar graph. Extract the letters from the graph to solve the riddle.

E = 1   M = 2   N = 3   T = 4   W = 5

What do you give a sick bird?

T W E E T M E N T
4 5 1 1 4 2 1 3 4

LESSON 7-4 Practice A
Reading and Interpreting Circle Graphs

The circle graph directly below shows the results of a survey of 50 teens. They were asked about their favorite fruits. Use the graph for Exercises 1–3.

1. Did more teens pick apples or grapes?

   apples

2. About what percent of teens picked strawberries?

   about 25%

3. According to the survey, 10% of teens chose oranges. How many teens chose oranges?

   5 teens

The circle graph below shows the results of a survey of 100 people. They were asked about their favorite doughnut flavors. Use the graph for Exercises 4–6.

4. Did more people pick frosted or filled?

   filled

5. About what percent of people picked glazed?

   about 50%

6. According to the survey, 30% of the people chose filled doughnuts. How many people chose filled?

   30 people

Would you use a bar graph or a circle graph to show the information? Explain your answer.

7. the number of bears that live in each of the national parks in Alaska

   a bar graph; One axis could show the number of bears, and the other axis could show the parks.

8. the number of visitors to Yellowstone National Park in August compared with the total number of summer visitors

   a circle graph; The circle represents all summer visitors.

Practice B
7-4 Reading and Interpreting Circle Graphs

The circle graph directly below shows the results of a survey of 80 teens who were asked about their favorite musical instruments. Use the graph for Exercises 1–3.

Favorite Musical Instruments

1. Did more teens pick piano or drums?
   piano
2. About what percent of teens picked guitar?
   about 25%
3. According to the survey, 20% of teens chose violin. How many teens chose violin?
   16 teens

The circle graph below shows the results of a survey of 100 people who were asked about their favorite vacation destinations. Use the graph for Exercises 4–6.

Favorite Vacation Destinations

4. Did more people pick mountains or beaches?
   beaches
5. About what percent of people picked mountains?
   about 25%
6. According to the survey, 15% of the people chose famous landmarks. How many people chose famous landmarks?
   15 people

Decide whether a bar graph or a circle graph would best display the information. Explain your answer.

7. number of tornadoes in each state during one year
   a bar graph; One axis could show the number of tornadoes and the other could show the states.

8. the number of pounds of Macintosh apples sold compared with the total number of pounds of apples sold at a market in one day
   a circle graph; The sector for the Macintosh apples could be compared to the entire circle.

Practice C
7-4 Reading and Interpreting Circle Graphs

The circle graph directly below shows the results of a survey of 60 teens who were asked about their favorite sports. Use the graph for Exercises 1–3.

Favorite Sports

1. Did more teens pick baseball or hockey?
   baseball
2. About what percent of teens picked tennis?
   about 10%
3. According to the survey, 40% of teens chose soccer. How many teens chose soccer?
   24 teens

The circle graph below shows the results of a survey of 200 people who were asked about their favorite breakfast foods. Use the graph for Exercises 4–6.

Favorite Breakfast Foods

4. Did more people pick cereal or eggs?
   cereal
5. About what percent of people picked toast?
   about 20%
6. According to the survey, 25% of the people chose pancakes. How many people chose pancakes?
   50 people

Decide whether a bar graph or a circle graph would best display the information. Explain your answer.

7. the amount of money budgeted for entertainment in one month compared with the total monthly allowance
   a circle graph; The sector for entertainment could be compared to the entire circle.

8. the number of calories different birds of prey need to consume in one day
   a bar graph; One axis could show the number of calories, and the other could show the birds.

Reteach
7-4 Reading and Interpreting Circle Graphs

A **circle graph** shows how the parts of a complete set of data are related. The circle shows 100% of the data.

Choose a circle graph to show what percent (or part) of the total sales is represented by each type of pizza.

Choose a bar graph to show the number of each type of pizza sold.

The circle graph shows that about one-half, or 50%, of the total sales was cheese pizzas. If 200 pizzas were sold, then about half, or 100, were cheese pizzas.

The circle graph shows the results of a survey of 100 teens who were asked about their favorite winter sport. Use the graph to answer each question.

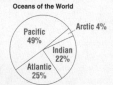
Favorite Winter Sport

1. Did more teens pick skiing or snowboarding?
   snowboarding
2. About what percent of teens picked skiing?
   about 25%
3. How many teens chose skiing?
   25 teens

Decide whether a bar graph or a circle graph would best display the information. Explain your answer.

4. the number of hurricanes in each state during 1 year
   bar graph; One axis could show the number of hurricanes, and the other could show the states.

5. the number of students who play soccer compared with the total number of students in a school
   circle graph; The sector for those who play soccer could be compared to the entire circle.

Challenge
7-4 Circle the Oceans

The table below shows the approximate percent of area that each ocean comprises of the total area of the oceans of the world. These oceans together actually make up one large, connected body of water.

Ocean	Approximate Area
Pacific	49%
Atlantic	25%
Indian	22%
Arctic	4%

1. Complete the circle graph at the right by labeling the ocean and percent of area for each part.

Oceans of the World
- Pacific 49%
- Arctic 4%
- Indian 22%
- Atlantic 25%

2. The oceans of the world cover almost 130 million square miles. About how many square miles are covered by the Pacific Ocean?
   about 64 million mi^2
3. About how many square miles are covered by the Atlantic Ocean?
   about 33 million mi^2
4. About how many square miles are covered by the Indian Ocean?
   about 29 million mi^2
5. About how many square miles are covered by the Arctic Ocean?
   about 5 million mi^2
6. About how many square miles are covered by the Pacific Ocean and the Atlantic Ocean combined?
   about 96 million mi^2
7. About how many more total square miles are covered by the Atlantic, Indian, and Arctic oceans than by the Pacific Ocean?
   about 3 million mi^2

LESSON 7-4 Problem Solving
Reading and Interpreting Circle Graphs

Write the correct answer.

1. A market research group conducted a survey of 100 sports car owners. The group learned that 50% of the car owners loved their cars. What part of the circle in a circle graph would be represented by that statistic?
$\frac{1}{2}$ of the circle

2. Juanita has 100 CDs. In her collection, 37 of the CDs are rock music, 25 are jazz, and 38 are country music. What part of the circle in a circle graph would represent the jazz CDs?
$\frac{1}{4}$ of the circle

3. Mr. Martin wanted to compare his monthly rent to his total income. Should he use a circle graph or a bar graph?
circle graph

4. Mr. Martin's rent has increased every year for the last 6 years. Should he use a circle graph or bar graph to show the yearly increase?
bar graph

Choose the letter for the best answer. Use the circle graph.

5. To which age group do most of the fitness club members belong?
A 18–20
B 70+
C 30–39
D 40–49

6. There are 100 members in a fitness club. How many members does the graph suggest will be between the ages of 18 and 39?
F 10
G 33
H 43
J 22

7. Which 2 age groups make up more than one-half the members?
A 18–29 and 30–39
B 30–39 and 40–49
C 40–49 and 50–59
D 18–29 and 70+

8. Which 2 age groups make up 3 times as many members as those who are between 60 and 69?
F 40–49 and 50–59
G 50–59 and 70+
H 30–39 and 18–29
J 18–29 and 50–59

LESSON 7-4 Reading Strategies
Reading a Graph

This circle graph gives you the whole picture of Molly's 60-minute exercise program. You can see how each part of her exercise program relates to the whole program.

Molly's 60-Minute Exercise Program

Answer the questions about the circle graph.

1. What does the graph show?
Molly's 60-minute exercise program

2. On which activity does Molly spend half of her time?
aerobics

3. How do you know?
Half of the circle is labeled "aerobics."

4. On which two activities does Molly spend the same amount of time?
toning exercises and stretching exercises

5. How do you know?
They are equal sections of the circle.

You can figure out how many minutes Molly spends doing each activity by looking at the graph.

6. What is the total number of minutes Molly spends doing her full exercise program? __60 minutes__

7. How many minutes does Molly spend doing aerobics? (She spends one-half of her exercise program doing aerobics.) __30 minutes__

8. How many minutes does Molly spend doing stretching exercises? (She spends one-fourth of her time stretching.) __15 minutes__

LESSON 7-4 Puzzles, Twisters & Teasers
Cow-abunga!

Doctor Digby did a study of people's eating habits. He asked 100 people what kinds of meat they prefer. Fifty people said they like all kinds of meat. Twenty-five people said they don't like any kind of meat. Five people said they prefer fish. Ten people said they prefer chicken. Ten people said they prefer beef. Show the results of Doctor Digby's study as a circle graph using the empty circle below.

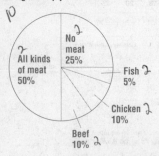

Use the key to answer the riddle: 5% = E 10% = D 25% = R 50% = U
Why did the cow cross the road?

To get to the __U__ __D__ __D__ __E__ __R__ side.
            50% 10% 10% 5%   25%

LESSON 7-5 Practice A
Box-and-Whisker Plots

1. Use the data to make a box-and-whisker plot.
24, 32, 35, 18, 20, 36, 12

The box-and-whisker plot shows the test scores of two students. Use the box-and-whisker plot for Exercises 2-4.

2. Which student has the greater median test score? __Amy__
3. Which student has the greater interquartile range of test scores? __Ed__
4. Which student has the greater range of test scores? __Ed__
5. Which student appears to have more predictable test scores? Explain your answer.
Amy; Possible answer: The range and interquartile range are smaller for Amy than for Ed, so Amy's test scores are more predictable.

The box-and-whisker plot shows prices of hotel rooms in two beach towns. Use the box-and whisker plot for Exercises 6-8.

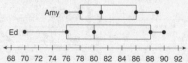

6. Which town has the greater median room price? __Port Eagle__
7. Which town has the greater interquartile range of room prices? __Port Eagle__
8. Which town appears to have more predictable room prices? Explain your answer.
Surfside; Possible answer: The interquartile range is smaller for Surfside than for Port Eagle, so Surfside's room prices are more predictable.

LESSON 7-5 Practice B
Box-and-Whisker Plots

1. Use the data to make a box-and-whisker plot.
19, 46, 37, 16, 24, 47, 23, 19, 31, 25, 42

Use the box-and-whisker plot of games won per season by the New York Yankees and the Arizona Diamondbacks for 1998-2005 for Exercises 2-4.

2. Which team has the greater median number of games won? __New York__
3. Which team has the greater interquartile range of games won? __Arizona__
4. Which team appears to have a more predictable performance? __New York__

Use the box-and-whisker plot of nightly tip totals that a waitress gets at two different restaurants for Exercises 5-7.

5. At which restaurant is the median tip total greater? __Greystone Inn__
6. At which restaurant is the interquartile range of tip totals greater?
__Greystone Inn__
7. At which restaurant does the tip total appear to be more predictable?
__Sam's Place__

LESSON 7-5 Practice C
Box-and-Whisker Plots

Use the data for Exercises 1-3.
38, 42, 26, 32, 40, 28, 36, 27, 29, 6, 30

1. Make two box-and-whisker plots of the data on the same number line: one plot with the outlier and one plot without the outlier.

2. How does the outlier affect the interquartile range of the data?
__It increases the interquartile range by 1.__
3. Which is affected more by the outlier: the range or the interquartile range?
__The range is more affected.__

The table shows scores for two golfers. Use the table for Exercises 4-7.

Henry	78	80	74	91	73	88	92	94	78	80
Trish	82	84	81	82	80	89	86	90	78	85

4. Make two box-and-whisker plots of the data on the same number line.

5. Which golfer has the lower median score? __Henry__
6. Which golfer has the lesser interquartile range of scores? __Trish__
7. Which golfer appears to be more consistent? __Trish__

LESSON 7-5 Reteach
Box-and-Whisker Plots

A **box-and-whisker plot** separates a set of data into four equal parts.

Use the data to create a box-and-whisker plot on the number line below:
35, 24, 25, 38, 31, 20, 27

1. Order the data from least to greatest.
__20 24 25 27 31 35 38__

2. Find the least value, the greatest value, and the median.
__20, 38, and 27__

3. The **lower quartile** is the median of the lower half of the data. The **upper quartile** is the median of the upper half of the data. Find the lower quartile and the upper quartile in this data.
__24, 35__

4. Above the number line, plot points for the least value, lower quartile, median, upper quartile, and greatest value.

5. Draw a box around the quartiles and the median. Draw a whisker from the least value to the lower quartile. Draw a whisker from the upper quartile to the greatest value.

Use the data to create a box-and-whisker plot:
63, 69, 61, 74, 78, 72, 68, 70, 65

6. Order the data. __61 63 65 68 69 70 72 74 78__
7. Find the least and greatest values. __61, 78__
8. Find the median, lower quartile, and upper quartile. __69, 64, and 73__
9. Plot the points, draw the box, and add the whiskers.

LESSON 7-5 Challenge
Puzzling Plots

The box-and-whisker plot below represents a data set that has 7 numbers. Use the box-and-whisker plot to find the numbers in the data set.

1. What are the least and greatest numbers in the data set? __10; 24__
2. What is the median? __18__
3. Is the median one of the 7 numbers in the data set, or is it the mean of the middle two numbers. Explain your answer.
__The median is a number in the data set. When there are an odd number of numbers, the median is the middle number.__
4. What are the lower and upper quartiles? __12, 23__
5. Are the lower and upper quartiles two of the numbers in the data set? Explain your answer.
__Yes. Since the lower and upper halves of the data each have an odd number of numbers, the median of each is the middle number in each.__
6. The remaining two numbers in the data set have a mean of 19 and range of 6. What are the two numbers? __16, 22__
7. The box-and-whisker plot below represents a data set that has 7 numbers. When the numbers are in order from least to greatest, the third and fifth numbers in the set have a mean of 12 and a range of 4. What are the 7 numbers?

__6, 9, 10, 12, 14, 15, 20__

LESSON 7-5 Problem Solving
Box-and-Whisker Plots

Write the correct answer.

A fitness center offers two different yoga classes. The attendance for each class for 12 sessions is represented in the box-and-whisker plot.

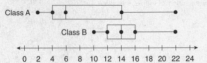

1. Which class has a greater median attendance? How much greater is it?
   Class B; 8

2. Which class appears to have a more predictable attendance?
   Class B

3. Which class has an attendance of less than 14 people 75% of the time?
   Class A

4. What percent of the time does Class B have an attendance greater than 16?
   25% of the time

Choose the letter for the best answer.

The box-and-whisker plot shows the percent of people in eight Central American countries who used the Internet in 2005.

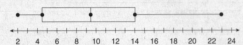

5. What is the range in the percents of people who used the Internet in the eight countries?
   A 23%   C 14%
   B 21%   D 9.5%

6. In how many of these eight countries did at least 9.5% of the people use the Internet?
   F less than 2   **H** at least 4
   G less than 4   J cannot be determined

7. What is the mode of the data?
   A 2%   C 9.5%
   B 4.5   **D** cannot be determined

8. What is the interquartile range of the data?
   F 23%   H 14%
   G 21%   **J** 9.5%

LESSON 7-5 Reading Strategies
Use Graphic Aids

A **box-and-whisker plot** shows a set of data divided into four equal parts called **quartiles**.

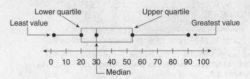

- The median score divides the set of data in half. The median score for this plot is 30.
- The box shows the middle half of the data, located on either side of the median. The box extends from 20 to 53.
- The two whiskers identify the remaining half of the data. One whisker extends from the box to the greatest value: from 53 to 90. The other whisker extends from the box to the least value: from 2 to 20.

Answer each question.

1. What does the box stand for in a box-and-whisker plot?
   the middle half of the data

2. How are the whiskers determined?
   The whiskers extend from both ends of the box to the least and greatest values.

3. Why is it important to find the median score?
   It divides the set of data in half.

Describe where these scores are located in the box-and-whisker plot above.

4. 18   between the **least value** and the **lower quartile**
5. 75   between the **upper quartile** and the **greatest value**
6. 45   between the **median** and the **upper quartile**

LESSON 7-5 Puzzles, Twisters & Teasers
The Plot Thickens!

Circle words from the word list in the word search below.

box   whisker   plot   first   third   upper
quartile   extreme   data   median   range   fraction

Now think about the alphabet and the letters you see above.

What two letters of the alphabet hold nothing when you say them together?

M   T

LESSON 7-6 Practice A
Line Graphs

Use the table to answer the questions.

U.S. Personal Spending on Selected Electronics

Year	Amount Spent ($billions, estimated)
1994	$71
1996	$80
1998	$90
2000	$107

1. Use the data in the table to make a line graph.

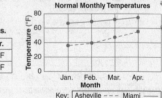

2. When did the amount spent on electronics increase the most?
   from 1998 to 2000

3. About how much was spent on electronics in 1999?
   Possible answer: about $100 billion

The table below shows normal monthly temperatures in Asheville, North Carolina, and Miami, Florida, for the first 4 months of the year. Use the table to answer the questions.

	Jan.	Feb.	Mar.	Apr.
Asheville	36°F	39°F	47°F	55°F
Miami	67°F	69°F	72°F	75°F

4. Make a double-line graph of the data.

5. During which month is there the greatest difference in temperatures between the two cities?
   during January

6. During which month is there the least difference in temperatures between the two cities?
   during April

7. In which city did the temperature vary the least during these 4 months?
   Miami

8. Where should you spend your April vacation if you want to go hiking wearing your new sweatshirt?
   Asheville

Practice B
7-6 Line Graphs

Use the table for Exercises 1–3.

Retail Price of Regular Gasoline in the United States (to the nearest cent)

Year	1990	1992	1994	1996	1998	2000	2002	2004
Price Per Gallon	$1.16	$1.13	$1.11	$1.23	$1.06	$1.51	$1.32	$1.82

1. Make a line graph of the data.

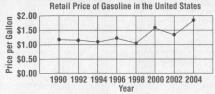

2. When did the cost of gasoline decrease the most?

between 2000 and 2002

3. About how much did gasoline cost in 1995?

Possible answer: about $1.17

The table below shows the student population at elementary schools in two cities, New City and Jackson.

Year	1996	1997	1998	1999	2000	2001	2002	2003
New City	450	460	440	430	495	500	600	645
Jackson	500	475	450	525	430	440	485	480

4. Make a double-line graph of the data.

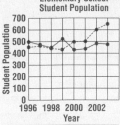

5. During which year did New City's school population increase the most?

from 2001 to 2002

6. The mall in Jackson closed. Many people lost their jobs and moved their families to New City, where a new mall opened. In what year did this probably happen? Explain your thinking.

2000; Possible answer: The student population jumped in New City, but dropped in Jackson.

Practice C
7-6 Line Graphs

Use the table for Exercises 1–3.

U.S. Population (to the nearest million)

Year	Number
1820	10
1840	17
1860	31
1880	50
1900	76
1920	106
1940	132
1960	179
1980	226
2000	281

1. Make a line graph of the data.

2. When did the U.S. population have the greatest increase?

from 1980 to 2000

3. Estimate the U.S. population in 1970.

about 200 million

The table below shows the price in cents per pound U.S. farmers received for apples and peanuts they grew.

Year	1996	1997	1998	1999	2000	2001	2002	2003
Apples (cents/lb)	20.8	22.1	17.1	21.3	17.8	22.9	25.6	29.5
Peanuts (cents/lb)	28.1	28.3	25.7	25.4	27.4	23.4	18.2	18.8

4. Make a double-line graph of the data in the table.

5. During which year was there the greatest difference in prices?

during 2003

6. During which year was there the least difference in prices?

during 2001

7. What trends do you see in the data?

Possible answer: From 2000-2003, the price of apples is increasing and the price of peanuts is decreasing.

Reteach
7-6 Line Graphs

A **line graph** shows data changing over time.
A **double-line graph** shows two sets of data changing over time.

1. The table shows the normal daily low temperature in °F in Indianapolis and San Francisco. Use the data to make a double-line graph.

Month	Jan.	Mar.	May	Jul.	Sept.	Nov.
Indianapolis	26°	41°	63°	75°	67°	43°
San Francisco	49°	53°	58°	63°	65°	55°

a. Finish labeling each axis.
   What is the vertical axis label?
   Temperature (F°)
   What is the horizontal axis label?
   Month
b. Plot each point for each set of data.
c. Connect the points. Use a dotted line to represent one set of data.
d. Give the graph a title and key.

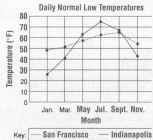

2. In which city was the temperature the most constant?

San Francisco

3. During which 2 months in Indianapolis did the temperature increase the most?

March to May

4. Use the graph to estimate the low temperature in October in San Francisco.

about 60°F

5. During which month was the difference in the daily normal low temperatures in Indianapolis and San Francisco most likely the smallest?

September

Challenge
7-6 What's My Line?

Write the letter of the graph that best represents each event described.

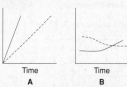

1. Two science students are timing ice melting and water freezing. Which graph shows the temperatures during both experiments?

B

2. In the music industry, when CDs became popular, sales of cassettes plummeted. Which graph shows this trend?

D

3. The batter takes a strike on the first pitch but hits a home run on the second pitch. Which graph shows the speed of the ball in each case?

C

4. Two brothers attend the same school. One walks and the other rides a bike. Which graph compares the time required for them to travel to school?

A

Draw a double-line graph to solve.
Possible answers are given.

5. The town swimming pool puts aside the last 15 minutes of every hour for an adult swim. Show the change in the number of children and adults in the pool over several hours.

6. The Ice Cream Station is open only during the summer months. Cool Guy's Ice Cream is open all year long. Show the change in ice cream sales over the course of a year for each business.

LESSON 7-6 Problem Solving
Line Graphs

Write the correct answer.

The line graph shows the number of households with cable television from 1996 to 2002.

1. About how many households had cable TV in 1998?
   about 67 million

2. About how many more households had cable TV in 2002 than in 1996?
   about 9 million

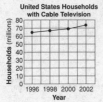
United States Households with Cable Television

3. During which two-year period did the number of households with cable TV grow the most?
   from 2000 to 2002

4. Use the graph to estimate the number of households with cable TV in 2001.
   about 71 million

Choose the letter for the best answer.
The double-line graph shows the number of tornadoes in the United States during part of 2002 and 2003.

5. About how many more tornadoes were there in August 2002 than in August 2003?
   A about 50   C about 30
   B about 40   D about 20

6. In which time period did the number of tornadoes increase both years?
   F Sept. to Oct.   H Nov. to Dec.
   G Oct. to Nov.   J none

7. Which time period showed the greatest decrease in the number of tornadoes?
   A Aug. to Sept. 2002
   B Aug. to Sept. 2003
   C Oct. to Nov. 2002
   D Nov. to Dec. 2003

8. During which month was there the greatest difference between the number of tornadoes in 2002 and in 2003?
   F Aug.   H Nov.
   G Sept.   **J** Dec.

LESSON 7-6 Reading Strategies
Organization Patterns

A **line graph** visually shows data that changes over time. Your height changes from year to year. Sales at a store change from month to month.

In the **line graph** below, look for patterns of change in a savings plan.

Brian's Savings Account

Answer the following questions about the line graph.

1. What information is located along the left side of the graph?
   number of dollars

2. In what intervals are the numbers shown on the left side of the graph?
   $5 intervals

3. What information is shown along the bottom of the graph?
   the months during which Brian saved money

Each point on the graph identifies the amount of money saved each month.

4. How much money had Brian saved in February?
   $15

5. How much had Brian saved in April?
   $15

6. Why do you think Brian's savings account dropped between February and March?
   Possible answer: He spent some of his savings.

LESSON 7-6 Puzzles, Twisters & Teasers
Rhyme Time TV!

Make a line graph of the data in the table. Think about the data and the way it is presented. Then make a rhyme to answer the riddle by filling in the letter associated with each ordered pair.

What do you call a table that shows information about cable TV?

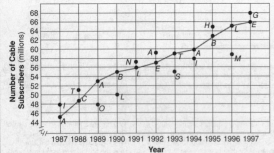

Year	Number of Cable Subscribers (millions)	Letter to Solve Riddle
1987	45	A
1988	49	C
1989	53	A
1990	55	B
1991	56	L
1992	57	E
1993	59	T
1994	60	A
1995	63	B
1996	65	L
1997	66	E

LESSON 7-7 Practice A
Choosing an Appropriate Display

Choose the letter of the graph that would best represent each type of data.

1. the number of visitors to a national park in each of the past five years
   A stem-and-leaf plot
   B Venn diagram
   C circle graph
   D line graph

2. the part of a whole group that said yes, no, or no opinion in a survey
   F stem-and-leaf plot
   G Venn diagram
   H circle graph
   J line plot

3. the distribution of heights of 30 boys in a gym class
   A stem-and-leaf plot
   B line graph
   C Venn diagram
   D circle graph

4. the number of students in each of four different clubs
   F circle graph
   G bar graph
   H line graph
   J Venn diagram

The table shows the distances four students ran during one week. Use the data in the table for Exercises 5 and 6.

Student	Miles Run
Janelle	15
Marty	25
Rosa	30
Sidney	10

5. Does the stem-and-leaf plot do a good job of representing the data from the table? Explain your answer.

Distances Run
Stems	Leaves
1	0 5
2	5
3	0

Key: 1|2 means 12 miles

Possible answer: No. There are only four values, and how often they occur and how they are distributed is not important.

6. Does the bar graph do a good job of representing the data? Explain.

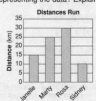
Distances Run

Yes. The bar graph shows how many miles each student ran, and makes it easy to compare the amounts.

LESSON 7-7 Practice B
Choosing an Appropriate Display

Choose the type of graph that would best represent this data.

1. the number of points scored by five different basketball players in a game

   possible answer: bar graph

2. the distribution of test scores in a math class

   possible answers: stem-and-leaf graph or line plot

3. the students who are in the chess club, the debating club, and the computer club, and the students who are in more than one of those clubs

   possible answer: Venn diagram

4. the percent of total income a family uses for rent, food, clothing, entertainment, savings and other expenditures

   possible answer: circle graph

The table shows the earnings of The Sandman Company during each of five years. Explain why each display does or does not appropriately represent the data.

Year	Earnings (millions of dollars)
2001	1
2002	3
2003	8
2004	12
2005	16

5.

   Earnings of The Sandman Company

   A circle graph displays data as parts of a whole. Since data in this table do not represent parts of whole, a circle graph is not appropriate.

6.

   Earnings of The Sandman Company

   The table shows data that changes over time. Since a line graph shows how data changes over time, this graph is appropriate.

LESSON 7-7 Practice C
Choosing an Appropriate Display

Choose the type of graph that would best represent this data.

1. the distribution of scores of a professional golfer for 20 rounds of golf

   possible answers: stem-and-leaf plot or line plot

2. the change in weight over time when the diet of a dog is changed.

   possible answers: line graph or bar graph

The table shows the responses of 100 students to a survey. Explain why each display does or does not appropriately represent the data.

Favorite Sport to Play	Number of Students
Baseball	15
Basketball	30
Football	10
Soccer	40
Tennis	5

3. Favorite Sport to Play

   A circle graph shows how data is divided into parts. Since this circle graph shows the part of the total number of votes that each sport received, it is appropriate.

4.

   A line plot shows how often data values occur and how they are distributed. Since the frequency and distribution of the data in the table is not important, this graph is not appropriate.

5. The data gives the heights in inches of 20 seventh-graders.
   58, 64, 60, 56, 66, 61, 60, 58, 70, 62, 60, 58, 56, 71, 62, 60, 61, 64, 59, 56
   Make the type of graph that would best represent the data.

   Check students' graphs. Possible answers: a line plot or a stem-and-leaf plot.

LESSON 7-7 Reteach
Choosing an Appropriate Display

There are different ways to show data. To choose a data display, think about the information that each kind of graph shows.

Bar graph: Compares amounts.

Circle graph: Shows data as parts of a whole.

Stem-and-leaf plot: Shows each value and how the values are distributed.

Line plot: Shows each value and how the values are distributed.

Line graph: Shows how amounts change over time.

Venn diagram: Shows members of sets and how sets overlap.

Circle the type of graph which would best show the data. Explain the reason for your choice. Possible answers are given.

1. the way the vote is split among three candidates in a school election
   (circle graph) or stem-and-leaf plot

   The vote is split into parts. A circle graph shows parts of a whole.

2. the average income for a family of four in 1970, 1980, 1990, and 2000
   line plot or (line graph)

   The data involves amounts changing over time. A line graph shows amounts changing over time.

3. the points scored by a football team during each game in a 16-game season
   (stem-and-leaf plot) or circle graph

   The display should show each point total. A stem-and-leaf plot shows each piece of data.

4. the calories per ounce in four different cuts of meat
   line plot or (bar graph)

   The display should compare the four pieces of data. A bar graph compares data.

LESSON 7-7 Challenge
Divided Bar Graphs

Suppose you wanted to show how parts of whole change over time. One way to do this would be to draw a series of circle graphs.

These circle graphs show the percent of TV households that have color TVs or only black and white TVs.

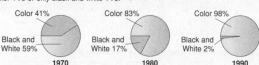

Another display you could use is a divided bar graph. In a divided bar graph, each bar is divided into sections that show parts of a whole. This divided bar graph represents the same data as the three circle graphs above.

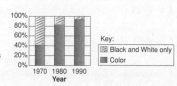

The table shows how many hours of television students at one high school watched. Use the data from the table to complete the divided bar graph.

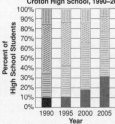

Television Viewing Habits at Croton High School, 1990-2005

	1990	1995	2000	2005
1 hour or less	40	44	70	120
1 to 2 hours	100	124	175	180
2 to 3 hours	200	176	80	50
More than 3 hours	60	56	50	30

Problem Solving
7-7 Choosing an Appropriate Display

Write the correct answer.

1. You take a survey of your class to find the number of years each student has lived in your town. You want to show how the data is distributed. What kind of graph would you use to display your data? Explain your choice.

   Possible answer: Line plot; a line plot shows each piece of data and how the data are distributed.

2. Wendy budgets $120 for the week. $30 is for transportation, $50 is for food, $25 is for entertainment, and $15 is for other expenses. What kind of graph would best represent the data? Justify your response.

   Possible answer: Circle graph; the parts of the budget represent parts of a whole, and a circle graph shows parts of a whole.

3. The New York Mets' worst season was their first season, in 1962. In that year they won 40 games. Their best season was in 1986, when they won 108 games. You want to show the number of games won each year in a way that makes it easy to see the distribution of the data. Why might you use a stem-and-leaf plot rather than a line plot?

   Possible answer: Since the data ranges from 40-108, a line plot would be very long.

4. During one school year, Matt reads 16 books, Tama reads 22 books, Rhonda reads 14 books, and Francisco reads 20 books. Would a line graph be an appropriate way to display this data? Explain your answer.

   Possible answer: No; a line graph displays changes over time, and this data does not involve time.

Choose the letter for the best answer.

5. Which type of graph would be most appropriate to show the distribution of daily high temperatures for a month?
   A circle graph
   B Venn diagram
   C stem-and-leaf plot
   D line graph

6. Which type of graph would be most appropriate to compare the price of the same TV at five discount stores?
   F stem-and-leaf plot
   G bar graph
   H circle graph
   J line plot

Reading Strategies
7-7 Use a Flowchart

You can use a flowchart to help you choose an appropriate display for a set of data.

Start Here
- Do you need to show amounts that change over time? Yes → Use a line graph.
- ↓ No
- Do you need to compare amounts? Yes → Use a bar graph.
- ↓ No
- Do you need to show parts of a whole? Yes → Use a circle graph.
- ↓ No
- Do you need to show two or more sets of data and how they overlap? Yes → Use a Venn diagram.
- ↓ No
- Do you need to see how often each piece of data occurs? Yes → Use a line plot or a stem-and-leaf plot.

Use the flowchart and the situation below for Exercises 1 and 2.

For a TV show, 55% of the viewers are under 18, 35% are between 18 and 34, and 10% are over 34. You want to show how big a part of the whole audience each age group is.

1. For which question in the flowchart is the answer yes?
   Do you need to show parts of a whole?

2. Which kind of graph will you use? circle graph

Use the flowchart and the situation below for Exercises 3 and 4.

A school has 34 different classes with sizes between 18 and 28 students. You want to show each of the different class sizes.

3. For which question in the flowchart is the answer yes?
   Do you need to see how often each piece of data occurs?

4. Which kind of graph will you use? line plot or stem-and-leaf plot

---

Puzzles, Twisters, & Teasers
7-7 Seeing Is Believing!

Choose the type of graph that would be best for each situation. Circle the letter above your answer.

1. comparing the numbers of raffle tickets sold by four students
   L line plot
   O bar graph

2. displaying every winning speed in the history of the Indianapolis 500 car race
   Y stem-and-leaf plot
   D circle graph

3. showing the factors and common factors of 24, 30, and 35
   A circle graph
   G Venn diagram

4. showing the sales of a monthly magazine in its first six months
   R Venn diagram
   E line graph

5. comparing the parts of a monthly budget
   D circle graph
   T stem-and-leaf plot

6. displaying quiz scores for 20 quizzes on which the scores can be from 1-10
   S line plot
   P line graph

Use the circled letters above the problem numbers to solve the riddle.

What do you need to spot an iceberg from 10 miles away?

G O O D   E Y E S
3. 1. 1. 5.   4. 2. 4. 6.

Practice A
7-8 Populations and Samples

Choose the letter of the sampling method that will better represent the whole population.

1. Clinton School Cafeteria: Student Satisfaction
   a. Mark surveys 40 students who are in his classes. 72% are satisfied with the food in the cafeteria.
   b. Tammy surveys 65 students by randomly choosing names from a list of all students in the school. 85% are satisfied with the food in the cafeteria.

2. Predicted Winner in an Election for Mayor
   a. Harris telephones 100 randomly chosen voters. 54% plan to vote for Mayor Green.
   b. Julia asks 70 people whom she knows. 45% plan to vote for Mayor Green.

For Problems 3-5, tell whether each sample may be biased. Explain your answer.

3. A town official surveys 50 people in a library to decide if town residents want the library expanded.
   The sample is biased. Possible explanation: It is likely that people who use the library will not have the same opinion as people who do not use the library.

4. A cable TV company randomly calls 200 customers and asks them if they are satisfied with their service.
   The sample is not biased. It is a random sample.

5. George surveys 15 students on the soccer team to learn whether middle school students want more money spent on school sports.
   The sample is biased. Possible explanation: It is likely that students who are not on sports teams will not have the same opinion as students who are on sports teams.

6. A factory produces 12,000 computers per week. The manager of a factory claims that fewer than 50 defective computers are produced each week. In a random sample of 500 computers, 2 were defective. Tell if the manager's claim is likely to be true.
   Yes; $\frac{2}{500} = \frac{48}{12,000}$, and $48 < 50$

Practice B
7-8 Populations and Samples

1. Determine which sampling method will better represent the entire population. Justify your answer.

Reading Habits of High School Students	
Sampling Method	Results of Survey
Dinah surveys 48 students who she knows.	91% have read a novel in the past month.
Suki gives survey forms to 100 students who were randomly chosen from a school attendance list.	59% have read a novel in the past month.

Suki's method will better represent the entire student population because she uses a random sample. Dinah's method will produce results that are not as representative of the entire population because she uses a convenience sample.

For Problems 2 and 3, determine whether each sample may be biased. Explain.

2. An on-line bookseller randomly chooses 200 book buyers from its database and then surveys those book buyers to find out if they were satisfied with the time it took to deliver their orders.

   The sample is not biased. It is a random sample.

3. Milena surveys 80 high school students who are leaving a jazz concert to determine the favorite type of music among high school students.

   The sample is biased. Possible explanation: It is likely that not all high school students will choose the same kind of music as the kind chosen by those who attend a jazz concert.

4. Zack chooses a random sample of 50 out of 400 students. He finds that 7 of them have traveled to a foreign country. Zack claims that over 50 of the 400 students have traveled to a foreign country. Do you agree? Explain your answer.

   Yes; $\frac{7}{50} = \frac{56}{400}$, and $56 > 50$

5. A mint produces 150,000 souvenir coins each year. In a random sample of 400 coins, 3 have a misprint. Predict the number of coins that will have misprints in a year.

   1,125 coins will have misprints.

Practice C
7-8 Populations and Samples

1. Determine which sampling method will better represent the entire population. Justify your answer.

The Midland Company: Employee Satisfaction	
Sampling Method	Results of Survey
Wanda interviews 80 employees who were randomly selected from the company payroll list.	69% feel adequately challenged by their jobs.
Bernard interviews the last 30 employees to be hired.	90% feel adequately challenged by their jobs.

Wanda's method will better represent the entire employee population because she uses a random sample. Bernard's method will produce results that are not as representative of the entire population because his sample is not random.

For Problems 2 and 3, determine whether each sample may be biased. Explain.

2. A landlord e-mails 60 of his 1,250 tenants and surveys them to determine whether they would like to use the Internet to pay rent.

   The sample is biased. Possible explanation: It is likely that not all tenants will give the same answer as those tenants who use e-mail.

3. An insurance company surveys 350 of its customers by randomly choosing names from its customer database and then telephoning the customers.

   The sample is not biased. It is a random sample.

Explain whether you would survey the entire population or use a sample.

4. You want to know how many hours members of a sports team train each week during the off-season.

   Possible answer: You would survey the entire population because the population is small.

5. You want to know the average income of people who eat at vegetarian restaurants across the country.

   Possible answer: You would use a sample because the entire population is large.

Reteach
7-8 Populations and Samples

Survey topic: number of books read by seventh-graders in Richmond

A **population** is the whole group that is being studied.	*Population*: all seventh-graders in Richmond
A **sample** is a part of the population.	*Sample*: all seventh graders at Jefferson Middle School
A **random sample** is a sample in which each member of the population has an equal chance of being chosen. A random sample is a better representation of a population than a non-random sample.	*Random sample*: Have a computer select every tenth name from an alphabetical list of each seventh-grader in Richmond.
A **biased sample** is a sample that does not truly represent a population.	*Biased sample*: all of the seventh graders in Richmond who are enrolled in honors English classes.

Tell if each sample is biased. Explain your answer.

1. An airline surveys passengers from a flight that is on time to determine if passengers on all flights are satisfied.

   The sample is biased. The passengers on one on-time flight are likely to feel differently about their flight than passengers on some other flights.

2. A newspaper randomly chooses 100 names from its subscriber database and then surveys those subscribers to find if they read the restaurant reviews.

   The sample is not biased. It is a random sample.

3. The manager of a bookstore sends a survey to 150 customers who were randomly selected from a customer list.

   The sample is not biased. It is a random sample.

4. A team of researchers surveys 200 people at a multiplex movie theater to find out how much money state residents spend on entertainment.

   The sample is biased. The people who go to movies are not likely to spend amounts of money on entertainment that are similar to the amounts spent by other people in the state.

Challenge
7-8 A Sample of Samples

Consider the three sampling methods described below.

Systematic sample: You randomly select an individual and then follow a pattern to select others in the sample. For example, choose a name from the first 50 names in a telephone book. Then choose every fiftieth name after that.

Stratified sample: A population is divided into subgroups, called strata, that contain similar individuals. For example, interview all the boys and girls at a movie. Boys and girls are two strata.

Cluster sample: A population is divided into sections and then a few of the sections are selected. All the members of those sections are chosen. For example, suppose a city has 50 voting districts. Five of those districts are selected, and every person in each district selected is polled.

Identify the type of sampling used as systematic, stratified, or cluster.

1. A teacher selects every third student in the class. — systematic
2. A teacher surveys all students from each of 5 randomly selected classes. — cluster
3. A principal selects 3 girls and 3 boys from each of 10 classes. — stratified
4. A teacher selects 10 students under 12 years old and 10 students over 12 years old. — stratified
5. A reporter interviews all students in each of 4 randomly selected schools. — cluster
6. A reporter interviews 20 men and 20 women. — stratified
7. Workers on an assembly line check every tenth tire in a tire manufacturing plant. — systematic
8. An advertising consultant surveys all members in each of 3 randomly selected fitness clubs from all the fitness clubs in the city. — cluster
9. A department store manager selects a customer from the first 20 on a customer list. Then she selects every twentieth customer after that. — systematic
10. Members of the school board survey 25 elementary school students, 25 middle school students, and 25 high school students. — stratified

Problem Solving
7-8 Populations and Samples

Write the correct answer.

1. Max wants to find out the exercise habits of local children. He plans to survey every third child he sees coming out of a sporting goods store. Max says his sample is not biased. Do you agree? Explain your answer.

   No; Not all children are likely to have the same exercise habits as children who go to a sporting goods store.

2. Ms. Constantine is choosing among three field trips for her two classes. She wants to determine which trip her students prefer. Should she survey the entire population or use a sample? Explain.

   Possible answer: She should survey the entire population, since it is relatively small.

3. A researcher catches 60 fish from different locations in a lake. He then tags the fish and puts them back in the lake. Two weeks later, the researcher catches 40 fish from the same locations. 8 of these 40 fish are tagged. Predict the number of fish in the lake.

   300 fish

4. A high school has 1,800 students. A random sample of 80 shows that 24 have cell phones. Predict the number of students in the high school who have cell phones.

   540 students

Choose the letter for the best answer.

5. The school board wants to study computer literacy among teachers. Which would represent a random sample of teachers?
   - A all high school math teachers
   - B teachers from the middle school whose name begins with N
   - C all male teachers
   - **D** every eighth teacher on an alphabetical list

6. In a random sample, 3 of 400 computer chips are found to be defective. Based on the sample, about how many chips out of 100,000 would you expect to be defective?
   - **F** 750
   - G 3,000
   - H 4,000
   - J cannot be determined

Reading Strategies
7-8 Compare and Contrast

To get information about issues, a survey is conducted. Surveys can be done in two different ways.
- **Population** — The entire group is surveyed.
- **Sample** — Part of the entire group is surveyed.

1. Compare the difference between collecting information from the population and collecting a sample.

   Population is everyone and a sample is part of the population.

There are two different types of samples.
- **Unbiased sample** — The sample represents the population.
- **Biased sample** — The sample does not represent the population.

2. What is the difference between an unbiased sample and a biased sample?

   An unbiased sample represents the population and a biased sample does not.

Mrs. Jones wants to know which sport 7th graders in the district like best. There are 7th graders in 6 different schools in the district. She can collect data in one of the following ways:

Population—Ask every 7th grade student at all 6 schools.
Unbiased sample—Ask every other 7th grader at 3 of the schools.
Biased sample—Ask 7th grade boys at 3 of the schools.

Write "unbiased sample" or "biased sample" to describe each survey.

3. A survey conducted at an ice cream store asked only mothers their favorite ice cream flavor.

   biased sample

4. A reporter asked every tenth person coming out of a theater how well they liked the movie.

   unbiased sample

5. A survey asked only girls to identify their favorite item on the school cafeteria menu.

   biased sample

Puzzles, Twisters, & Teasers
7-8 Answer This!

Choose the best description of each sample. Circle the letter above your answer.

1. Tyler surveys 60 people at a high school football game to find out which high school sports team is most popular.

   O biased   A not biased

2. A computer manufacturer randomly calls 150 people who called its help line and surveys them to determine if they were satisfied with the help they received.

   M biased   **N** not biased

3. Margie surveys 12 of her friends to determine what percentage of students participate in after-school clubs.

   F random sample   **D** convenience sample

4. Tamira surveys 50 riders on one commuter train to determine whether riders on all trains in the system think that the on-time performance is adequate.

   K biased   A not biased

5. Sam calls every sixth student on a high school attendance list and surveys the students to determine their favorite subject.

   M random sample   R convenience sample

Write the circled letters above the problem numbers to solve the riddle.

What is the difference between dinosaurs and dragons?

Dinosaurs D O N O T S M O K E
              3.   1.   2.   1.   5.   1.   4.

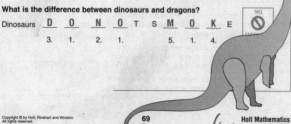

Practice A
7-9 Scatter Plots

The table shows how much energy was produced from wind power in the United States from 2000 to 2004.

Year	2000	2001	2002	2003	2004
Wind Power (trillions of BTUs)	57	70	105	115	143

1. Make a scatterplot of the data.

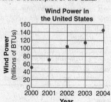

2. What does the scatter plot show about how much wind power was produced in the United States from 2001 to 2004?

   The amount of wind power increased over time.

3. What kind of correlation does the scatter plot show—positive, negative, or no correlation?

   positive correlation

Write *positive*, *negative*, or *no correlation* to describe each relationship.

4.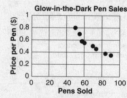

   negative correlation

5. Shoe Size Versus Golf Score

   no correlation

6. outside temperature and the chance of your getting frostbite

   negative correlation

7. number of hours of spent studying for a test and the test scores

   positive correlation

8. number of basketballs going into the basket and distance in feet from the basket when shooting

   negative correlation

9. height and age of adults over 25

   no correlation

LESSON 7-9 Practice B
Scatter Plots

The table shows boys' average heights in inches from ages 6 through 13. Use the table for Exercises 1–3.

Age	6	7	8	9	10	11	12	13
Height (in.)	$46\frac{3}{4}$	49	51	$53\frac{1}{4}$	$55\frac{1}{4}$	$57\frac{1}{4}$	59	61

1. Make a scatter plot of the data.

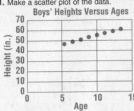

2. Describe the relationship between the data sets.

   As boys get older, they grow taller.

3. What kind of correlation does the plot show?

   positive correlation

Write *positive*, *negative*, or *no correlation* to describe each relationship.

4.

   negative correlation

5.

   positive correlation

6. student test scores and the number of students who walk to school

   no correlation

7. the grade levels of students and their ages in months

   positive correlation

8. the year a state entered the union and the number of years as a state

   negative correlation

9. ages of students and their grades on tests

   no correlation

LESSON 7-9 Practice C
Scatter Plots

The table shows the percentage of homes with a PC based on household income in 2001. Use the table for Exercises 1–3.

1. Make a scatter plot of the data.

Annual Income	Homes with PCs
Less than $5,000	26%
$5,000–$9,999	19%
$10,000–$14,999	26%
$15,000–$19,999	32%
$20,000–$24,999	40%
$25,000–$34,999	50%
$35,000–$49,999	64%
$50,000–$74,999	78%
$75,000 and up	89%

2. Describe the relationship between the data sets.

   The higher the household income, the greater the percentage of homes with a PC.

3. What kind of a correlation does the plot show?   positive

Write *positive*, *negative*, or *no correlation* to describe each relationship.

4.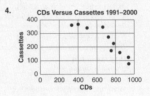

   negative correlation

5.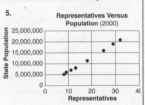

   positive correlation

6. the number of senators and the state population

   no correlation

7. the number of hours spent exercising and the number of calories burned

   positive correlation

CHAPTER 7-9 Reteach
Scatter Plots

A **scatter plot** can be used to explore how two sets of data are related.

A scatter plot can show three different relationships between data sets.

Positive Correlation — y-axis values increase as x-axis values increase.

Negative Correlation — y-axis values decrease as x-axis values increase.

No Correlation

Look at each scatter plot. What kind of correlation is shown?

1. Feeding a Family

   positive correlation

2. Science and Your Feet

   no correlation

What type of correlation might you expect between:

3. the number of pages in a book and the number of copies the book sells?

   no correlation

4. the number of computers per school and the number of students per computer?

   negative correlation

5. the number of students in a city and the number of students who play team sports?

   positive correlation

6. the temperature outside and the amount of heat used inside?

   negative correlation

7. a person's height and his or her birthday?

   no correlation

8. the speed of a car and the time required to travel a certain distance?

   negative correlation

LESSON 7-9 Challenge
Time On Task

Use a stopwatch to find out how long it takes to write your first and last names backwards. Do it 10 times, filling in the table with the time of each try. Use the data to make a scatter plot. **Answers will vary.**

Try	Time
1	
2	
3	
4	
5	
6	
7	
8	
9	
10	

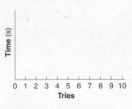

1. Does your scatter plot show positive, negative, or no correlation?

   Possible answer: negative

2. Summarize the data from the scatter plot.

   Possible answer: The more times I practiced writing my name backwards, the less time it took.

3. Check the tasks that would probably have a scatter plot that is similar to the one you drew above after doing each task 10 times.

Making tacos in a fast food restaurant		Writing multiplication facts from 1 to 12	✓
Filling boxes on an assembly line	✓	Eating dinner	
Driving from home to school		Talking to a friend on the phone	

4. Based on the data and scatter plot, what conclusion can you make about learning to do a job?

   Possible answer: The more you practice, the better and faster you will do the job.

Holt Mathematics

LESSON 7-9 Problem Solving
Scatter Plots

Write the correct answer.

This scatter plot compares the mean annual income of Americans with the number of years spent in school.

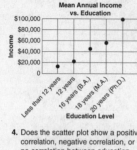

1. Which level of education has a mean annual income between $40,000 and $50,000?

   16 years, B.A.

2. Estimate the range of income data on this scatter plot.

   about $85,000

3. Which level of education has the lowest income?

   less than 12 years

4. Does the scatter plot show a positive correlation, negative correlation, or no correlation between education and income?

   positive correlation

Choose the letter for the best answer.

5. What kind of correlation would you expect to find between a city's annual snowfall amount and the size of its population?
   A positive correlation
   B negative correlation
   C no correlation
   D impossible to say

6. What kind of correlation would you expect to find between a movie's length and the number of times it can be shown in a day?
   F positive correlation
   G negative correlation
   H no correlation
   J impossible to say

7. What kind of correlation would you expect to find between an animal's mass and the number of calories it consumes in a day?
   A positive correlation
   B negative correlation
   C no correlation
   D impossible to say

8. What kind of correlation would you expect to find between a person's height and his or her income?
   F positive correlation
   G negative correlations
   H no correlation
   J impossible to say

LESSON 7-9 Reading Strategies
Drawing Conclusions

You can place individual data points on a graph to see if a pattern occurs. This type of graph is called a **scatter plot**.

Scatter Plot A identifies a car's age and its mileage. Age in years is listed along the bottom of the graph. Mileage is listed along the left side of the graph in intervals of 10,000 miles.

Notice how the points for each data pair create a graph with an upward pattern. This indicates a positive relationship between the number of miles a car has been driven and the age of the car.

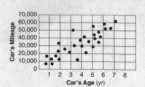

Scatter Plot B identifies a car's value and its mileage. Miles are given in intervals of 10,000. Values are in intervals of $5,000.

Notice how the points for each data pair create a downward pattern. This shows that the greater the mileage, the less a car is worth.

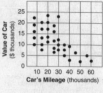

Scatter Plot C identifies a car's age and the number of passengers in the car.

Notice how the points are scattered on the graph. This shows that there is no relationship between these two sets of data.

Read each statement. Identify the scatter plot that helps you determine the answer, and write "true" or "false."   *2 each*

1. The older the car, the fewer the miles. __**Plot A; false**__

2. The more miles on a car, the less its value. __**Plot B; true**__

3. The newer the car, the fewer the miles.
   __**Plot A; true**__

4. The older the car, the more passengers it can carry. __**Plot C; false**__

LESSON 7-9 Puzzles, Twisters & Teasers
Scatter Brained!

Does the size of the brain determine the size of the body? Chart the data from the table below as a scatter plot. Then answer the questions.

Animal	Brain Weight (oz)	Body Weight (lb)
Human	53	140
Bottlenose Dolphin	56	340
Asian Elephant	263	10,000
Killer Whale	197	12,000
Cow	17	1,000
Mouse	0.014	0.024

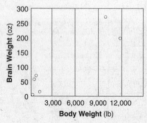

1. What kind of correlation do you see in the scatter plot?
   A positive correlation
   B negative correlation
   C no correlation

2. A mouse's brain makes up 3.2% of its body weight. A human's brain makes up 2.1% of its body weight. A cow's brain makes up 0.1% of its body weight. Which animal has the biggest brain in relation to its size?
   A mouse
   B human
   C cow

3. Which animal is the smartest, a cow, a human, or a bottlenose dolphin?
   A cow
   B human
   C bottlenose dolphin

4. Based on your conclusions, does a bigger brain make a bigger body?
   A can't tell
   B no
   C sure

Use the underlined letters to solve this brain teaser.
How do you count a herd of cows?

With a **C O W C U L A T O R**

LESSON 7-10 Practice A
Misleading Graphs

1. Which graph could be misleading? Why?

   Graph A   Graph B

   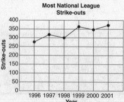 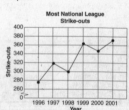

   Graph B; Possible answer: The vertical axis is broken; it looks as if the number of strikeouts in 2001 is 10 times as great as in 1996. (3)

Tell why each graph could be misleading.

2.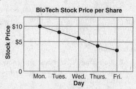

3.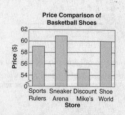

The intervals on the vertical axis are unequal.

The broken axis makes Discount Mike's look much less expensive.

LESSON 7-10 Practice B
Misleading Graphs

1. Which graph could be misleading? Why?

Graph A 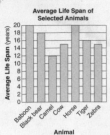 Graph B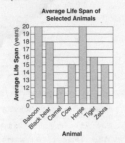

Graph B; Possible answer: The vertical axis is broken; it looks as if the camel lives about $\frac{1}{5}$ as long as a horse, but that is not true.

Explain why each graph could be misleading.

2.

3.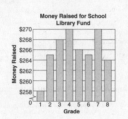

The intervals are too great to show the difference in speeds.

The broken axis makes it look as if there is a large difference in the amount raised.

LESSON 7-10 Practice C
Misleading Graphs

1. Which graph could be misleading? Why?

Graph A 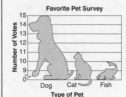 Graph B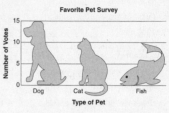

Graph A: Possible answer: The drawing of the dog is so large that it looks as if the dog got twice as many votes as any other animal.

Explain why this graph could be misleading. Then redraw it so it is not misleading.

2.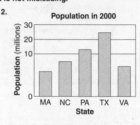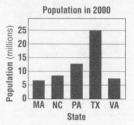

Unequal intervals make the populations look similar.

LESSON 7-10 Reteach
Misleading Graphs

Here are some ways a graph may be misleading.
- An axis is "broken" or numbers are left out on an axis.
- The intervals on an axis are not the same length.
- Different sizes are used to represent bars in a bar graph.
- Pictorial graphs could distort the data.

Compare these graphs.

It looks like sales dropped by more than half from January to February. Look at the vertical axis. It is broken, so the scale does not start at 0. This graph is misleading.

Look at the scale on the vertical axis. All the numbers from 0 to 1,000 are represented. This is a more realistic graph.

Explain why each graph is misleading.

1.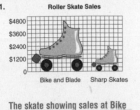

2.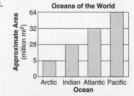

The skate showing sales at Bike and Blade is large, so it looks like the sales were more than twice as much.

The intervals on the vertical axis are not the same length.

LESSON 7-10 Challenge
Picture the Headlines

Read each fact. Draw a bar or line graph to support the two different headlines that report the fact.

1. Fact: In 2002, property taxes were $2.50 per $100 of the value of a house in Oakville. In 2003, they will increase to $2.60 for each $100 of value. **Possible answers are given.**

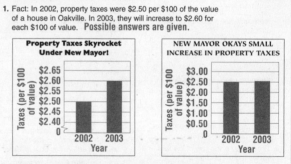

2. Fact: From 1996 to 1997, public schools in the United States averaged 7.8 students per computer. From 1997 to 1998, this average went down to 6.1. From 1998 to 1999, it was 5.7; and from 1999 to 2000, it was 5.4

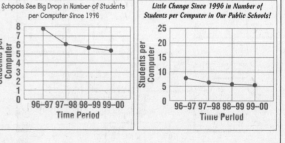

LESSON 7-10 Problem Solving
Misleading Graphs

Write the correct answer. Use the line graph for Exercises 1–4.

1. What would be a less misleading title for this graph?

   Possible answer: Spending, 1990–2000

2. How is the horizontal axis misleading?

   The intervals between years are not equal.

3. How is the vertical axis misleading?

   broken scale

4. How much did spending really increase between 1993 and 2000?

   about $40,000

Choose the letter for the best answer.

The bar graph is an advertisement used by a tour company to convince New Yorkers to vacation in Hawaii.

5. How far is New York from Los Angeles?
   - (A) 2,500 miles
   - B 4,000 miles
   - C 5,000 miles
   - D 5,200 miles

6. How far is New York from Hawaii?
   - F 2,500 miles
   - G 2,600 miles
   - (H) 5,000 miles
   - J 6,000 miles

7. What is the point of the ad?
   - A New York is closer to Hawaii than to Los Angeles.
   - B Hawaii is the same distance from New York as Los Angeles.
   - (C) Hawaii is only slightly farther from New York than is Los Angeles.
   - D San Francisco and Los Angeles are the same distance from New York.

8. Why is the graph misleading?
   - F The distances are incorrect.
   - G The bars are mislabeled.
   - H The bars are too tall.
   - (J) The intervals on the vertical axis are not equal.

LESSON 7-10 Reading Strategies
Analyze Information

Graphs can be misleading if the data is not shown properly.

Answer the following questions about Graphs A and B.

1. What number does the scale start on for Graph A? ___8___

2. What number does the scale start on for Graph B? ___0___

3. Which of the graphs is misleading? Why?

   Graph A; Because of the size of the bars in Graph A, it looks like apples are almost twice as popular as oranges.

A graph is misleading if the intervals are too large or too small.

Answer the following questions about Graphs C and D.

4. What intervals are used in Graph C? ___100___

5. What intervals are used in Graph D? ___300___

6. Which graph is misleading? Why?

   Graph D; There appears to be little difference between taco and spaghetti dinners sold.

LESSON 7-10 Puzzles, Twisters & Teasers
How's Your Vocabulary?

Complete the crossword puzzle by answering the clues.

Across

2. Each _____ or slice of a circle represents one part of the entire data set.
5. Each point on a scatter _____ represents a pair of data values.
7. bar or line _____
8. Frequency and cumulative frequency are two types of _____.
11. The _____ is the value or values that occur most often in a data set.
12. A _____ graph shows change over time for two sets of data.
13. The _____ is the sum of the data values divided by the number of data items.
14. The difference between the least and greatest values is called the _____.

Down

1. Often researchers can't survey every member of a large group, so they study a part of the group. This group is called a _____.
3. Positive, negative, or no _____ are three ways to describe data in a scatter plot.
4. the singular form of data
6. An extreme value in a set is called an _____.
9. The _____ is the middle value of an odd number of items arranged in order and it is the average of the two middle values for an even number of items.
10. A _____ table is one way to organize data into categories or groups.

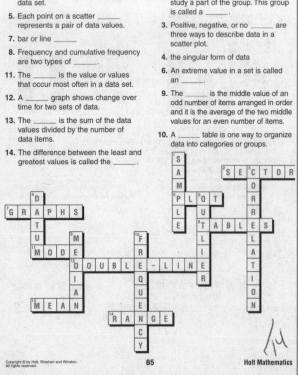